I0010113

Edge Computing with Amazon Web Services

A practical guide to architecting secure edge cloud infrastructure with AWS

Sean Howard

Edge Computing with Amazon Web Services

Copyright © 2024 Packt Publishing

All rights reserved. No part of this book may be reproduced, stored in a retrieval system, or transmitted in any form or by any means, without the prior written permission of the publisher, except in the case of brief quotations embedded in critical articles or reviews.

Every effort has been made in the preparation of this book to ensure the accuracy of the information presented. However, the information contained in this book is sold without warranty, either express or implied. Neither the author, nor Packt Publishing or its dealers and distributors, will be held liable for any damages caused or alleged to have been caused directly or indirectly by this book.

Packt Publishing has endeavored to provide trademark information about all of the companies and products mentioned in this book by the appropriate use of capitals. However, Packt Publishing cannot guarantee the accuracy of this information.

Associate Group Product Manager: Preet Ahuja

Publishing Product Manager: Surbhi Suman

Book Project Manager: Ashwini Gowda

Senior Editor: Divya Vijayan

Technical Editor: Arjun Varma

Copy Editor: Safis Editing

Proofreader: Safis Editing

Indexer: Rekha Nair

Production Designer: Vijay Kamble

Senior DevRel Marketing Executive: Linda Pearlson

DevRel Marketing Coordinator: Rohan Dobhal

First published: February 2024

Production reference: 1020224

Published by Packt Publishing Ltd.

Grosvenor House

11 St Paul's Square

Birmingham

B3 1RB, UK

ISBN 978-1-83508-108-2

www.packtpub.com

To Nick, Alice, and Angelica, thanks for putting up with this.

– Sean Howard

Contributors

About the author

Sean Howard is a principal **solutions architect** (**SA**) at **Amazon Web Services** (**AWS**). He's held several roles at AWS, including partner SA, SA manager, and now, principal SA, supporting aerospace and satellite customers in EMEA. Before AWS, he spent seven years at VMware as a network virtualization specialist. Before that, he spent a number of years in operations at the worlds second largest DNS registrar. His introduction to IT came during his time in the US Marines, where he worked on shipboard and field-deployed networks.

He holds a bachelor of science in computer information systems from Excelsior College, as well as a number of technology certifications, such as the VMware VCDX (#130).

Special contribution

Jay Naves supports aerospace and satellite customers at AWS with edge computing in space. In this role, he led the technical deployment of AWS Snowcone to the **International Space Station** (**ISS**) for use as an edge compute device and, at the time of writing, continues to manage its operation.

He is an expert in Linux systems administration and, prior to joining AWS, spent 20 years creating and leading **high-performance compute** (**HPC**), systems, and security teams at General Dynamics IT.

Jay holds the Red Hat Certified System Administrator and AWS Professional certifications.

About the reviewer

Maurice "Mo" Ramsey is an AWS disaster response and humanitarian leader. In this role, Mo brings significant business and technology leadership experience, advancing cloud technology adoption to support, enable, and activate **humanitarian assistance and disaster response (HADR)** missions for customers, partners, and communities.

Previously, Mo served in the US Army (active) from 1992 to 1997. He was honorably discharged in 1997 as a disabled veteran. He has served in leadership roles such as director at Slalom Consulting over advanced infrastructure and strategy (serving as a CIO "on-demand" resource), senior director for the cloud at CenturyLink (formerly Tier3), GM of advisory services at Lighthouse, and, most recently, area practice lead of professional services for nonprofits with AWS.

Mo earned a BA from Columbia College and holds multiple business and technology certifications.

About the reviewers

Table of Contents

3

Understanding Network and Security for Far-Edge Computing   43

Part 2: Introducing AWS Edge Computing Services

4

5

6

7

Using AWS Wavelength Zones on Public 5G Networks   169

Part 3: Building Distributed Edge Architectures with AWS Edge Computing Services

8

Utilizing the Capabilities of the AWS Global Network at the Near Edge   193

9

Architecting for Disconnected Edge Computing Scenarios   213

10

Utilizing Public 5G Networks for Multi-Access Edge (MEC) Architectures   237

11

Addressing the Requirements of Immersive Experiences with AWS   251

Part 4: Implementing Edge Computing Solutions via Hands-On Examples and More

12

Configuring an AWS Snowcone Device to Be an IOT Gateway   273

13

Deploying a Distributed Edge Computing Application   305

14

Preface

During the last few decades, computing models have fluctuated between drives toward centralization and distribution. At first, all compute horsepower was centralized on a mainframe, user terminals had no ability to operate independently, and networking was very simple. Then, client-server architectures took over, and corporate data centers were filled with racks full of servers that weren't much different than the clients they served. Next, those servers were virtualized to improve utilization and manageability. Finally, the bulk of those VMs were centralized in the cloud for similar reasons. Now, we are facing demand for a new hybrid model of cloud operation – distributed edge computing, sometimes known as Industry 4.0 or hybrid edge.

This shift is driven by the incredible growth in the number of networked devices, the amount of data they produce, and advances in our ability to process this data. A new hybrid edge computing model has emerged. This book aims to demystify this emerging computing model, particularly through the lens of architecting distributed edge applications in the **Amazon Web Services (AWS)** cloud.

Distributed edge computing allows us to process data and make decisions closer to its source. This approach reduces response time, lowers cost, and supports new use cases. AWS, with its world-class cloud and history of innovation, is uniquely positioned to help you capitalize on the strengths of centralization and distribution.

AWS services facilitate the deployment, management, and hosting of application components wherever they are needed. The AWS strategy is about reimagining what computing can look like when it is not constrained by physical location.

Throughout this book, we will explore offerings such as AWS Outposts, AWS Snow Family, AWS Wavelength, AWS Local Zones, AWS IoT Greengrass, and Amazon CloudFront. Whether it's processing data on a remote oil rig, processing and distributing live video, supporting the latency requirements of augmented reality, or running a full-scale data center in a disconnected environment, AWS has a solution.

After covering the *what* and *why* of distributed edge computing, this book explains the *how* – with hands-on exercises and an example of **Infrastructure-as-Code (IaC)** that you can use as a starting point for your own applications.

Who this book is for

This book is for cloud architects, cloud engineers, solution architects, and enterprise architects who are already familiar with the AWS cloud. It is particularly useful for those facing requirements to move compute, storage, database, and/or machine learning resources closer to the sources of data – but who aren't sure how to do so in a scalable, secure, and cost-effective manner.

What this book covers

Chapter 1, Getting Started with Edge Computing on AWS, introduces the concept of edge computing and its integration with cloud computing with a focus on AWS's strategy and tools for edge computing solutions.

Chapter 2, Understanding Network and Security for Near Edge Computing, dives into the challenges and solutions for networking and security specific to near edge computing, including private WANs, GSLB, IP Anycast, and new protocols such as HTTP/3 and QUIC.

Chapter 3, Understanding Network and Security for Far Edge Computing, covers the networking and security aspects of far edge computing, including RF communications, cellular networks, Wi-Fi connectivity, Low-powered networks such as LoRaWAN, and integration with satellite communication systems (SATCOM).

Chapter 4, Addressing Disconnected Scenarios with AWS Snow Family, introduces AWS's Snow Family products (Snowball Edge and Snowcone) and how they address the needs of disconnected or remote computing scenarios.

Chapter 5, Incorporating AWS Outposts into Your On-Premise Data Center, introduces AWS Outposts, offering insights into how it integrates with on-premise data centers and the various deployment options such as Outposts Rack and Server.

Chapter 6, Lowering First-Hop Latency with AWS Local Zones, introduces AWS Local Zones, explaining how they reduce latency by connecting on-premise networks to local AWS resources and routing internet traffic effectively into region-based applications.

Chapter 7, Using AWS Wavelength Zones on Public 5G Networks, introduces AWS Wavelength Zones, exploring their role in 5G networks, VPC extension, and integration with other AWS services.

Chapter 8, Utilizing the Capabilities of the AWS Global Network at the Near Edge, provides an overview of the AWS Global Network, its role in edge computing, and specific services offered at its edge such as Amazon CloudFront and AWS Global Accelerator.

Chapter 9, Architecting for Disconnected Edge Computing Scenarios, focuses on designing solutions for environments with limited connectivity, discussing AWS IoT services, tactical edge scenarios, and private 5G networks.

Chapter 10, Utilizing Public 5G Networks for Multi-Access Edge (MEC)Architectures, covers the architecture of 5G-based MEC solutions, comparing Wi-Fi and 5G in terms of observability, security, and capacity, and discusses applications such as V2X and software-defined video production.

Chapter 11, Addressing the Requirements of Immersive Experiences with AWS, discusses creating immersive experiences using AWS, including applications in online gaming, connected workers, and augmented/virtual reality.

Chapter 12, Configuring an AWS Snowcone Device to Be an IoT Gateway, details the process of setting up an AWS Snowcone as an IoT gateway, from ordering and configuring the device to deploying backend services and IoT Greengrass with AWS CloudFormation.

Chapter 13, Deploying a Distributed Edge Computing Application, details the process of quickly pushing a containerized application that runs on AWS Elastic Kubernetes Service, which has elements in a core Region, AWS Local Zones, and AWS Wavelength.

Chapter 14, Preparing for the Future of Edge Computing with AWS, concludes the book by discussing the future trends in edge computing, including business drivers, foundational strategies, and emerging patterns and anti-patterns in this field.

To get the most out of this book

It is assumed that you have a basic familiarity with general IT concepts such as IP networking, virtual machines, servers, and data centers, as well as an understanding of common AWS services equivalent to the AWS Certified Solution Architect – Associate level.

Software/hardware covered in the book	Operating system requirements
AWS **Command-Line Interface (CLI)** >= 2.13.9	Windows, macOS, or Linux
AWS Snowball Edge Client >= 1.2	Windows, macOS, or Linux
AWS OpsHub >= 1.15	Windows, macOS, or Linux
HashiCorp Terraform >=1.6	Windows, macOS, or Linux
Kubectl >= 1.28	Windows, macOS, or Linux

If your workstation is Windows-based, it is strongly recommended that you install **Windows Subsystem for Linux** (**WSL**), specifically the Ubuntu LTS 22(x) environment available from the Windows Store. This will allow you to directly use the example commands given with no modification.

If you are using the digital version of this book, we advise you to type the code yourself or access the code from the book's GitHub repository (a link is available in the next section). Doing so will help you avoid any potential errors related to the copying and pasting of code.

Download the example code files

You can download the example code files for this book from GitHub at `https://github.com/PacktPublishing/Edge-Computing-with-Amazon-Web-Services`. If there's an update to the code, it will be updated in the GitHub repository.

We also have other code bundles from our rich catalog of books and videos available at `https://github.com/PacktPublishing/`. Check them out!

Code in Action

The Code in Action videos for this book can be viewed at `https://bit.ly/3Ued07P`

Conventions used

There are a number of text conventions used throughout this book.

`Code in text`: Indicates code words in text, database table names, folder names, filenames, file extensions, pathnames, dummy URLs, user input, and Twitter handles. Here is an example: "Mount the downloaded `WebStorm-10*.dmg` disk image file as another disk in your system."

When we wish to draw your attention to a particular part of a code block, the relevant lines or items are set in bold:

```
[default]
exten => s,1,Dial(Zap/1|30)
exten => s,2,Voicemail(u100)
exten => s,102,Voicemail(b100)
exten => i,1,Voicemail(s0)
```

Any command-line input or output is written as follows:

```
aws configure \
set aws_access_key_id "your_access_key"
aws configure \
set aws_secret_access_key "your_secret_key"
aws configure \
set region "your_region"
```

Bold: Indicates a new term, an important word, or words that you see onscreen. For instance, words in menus or dialog boxes appear in **bold**. Here is an example: "Select **System info** from the **Administration** panel."

> **Tips or important notes**
> Appear like this.

Get in touch

Feedback from our readers is always welcome.

General feedback: If you have questions about any aspect of this book, email us at `customercare@packtpub.com` and mention the book title in the subject of your message.

Errata: Although we have taken every care to ensure the accuracy of our content, mistakes do happen. If you have found a mistake in this book, we would be grateful if you would report this to us. Please visit `www.packtpub.com/support/errata` and fill in the form.

Piracy: If you come across any illegal copies of our works in any form on the internet, we would be grateful if you would provide us with the location address or website name. Please contact us at `copyright@packt.com` with a link to the material.

If you are interested in becoming an author: If there is a topic that you have expertise in and you are interested in either writing or contributing to a book, please visit `authors.packtpub.com`.

Share Your Thoughts

Once you've read *Edge Computing with Amazon Web Services*, we'd love to hear your thoughts! Scan the QR code below to go straight to the Amazon review page for this book and share your feedback.

https://packt.link/r/1835081088

Your review is important to us and the tech community and will help us make sure we're delivering excellent quality content.

Download a free PDF copy of this book

Thanks for purchasing this book!

Do you like to read on the go but are unable to carry your print books everywhere?

Is your eBook purchase not compatible with the device of your choice?

Don't worry, now with every Packt book you get a DRM-free PDF version of that book at no cost.

Read anywhere, any place, on any device. Search, copy, and paste code from your favorite technical books directly into your application.

The perks don't stop there, you can get exclusive access to discounts, newsletters, and great free content in your inbox daily

Follow these simple steps to get the benefits:

1. Scan the QR code or visit the link below

https://packt.link/free-ebook/9781835081082

2. Submit your proof of purchase

3. That's it! We'll send your free PDF and other benefits to your email directly

Part 1:
Compute, Network, and Security Services at the Edge

Part One provides an introduction to distributed edge computing. It focuses on specific challenges and solutions related to networking and security in such architectures. This section is essential for understanding why you would want to use distributed edge computing as well as how you would go about managing data and network traffic efficiently and securely at the edge.

This part has the following chapters:

- *Chapter 1, Getting Started with Edge Computing on AWS*
- *Chapter 2, Understanding Network and Security for Near Edge Computing*
- *Chapter 3, Understanding Network and Security for Far Edge Computing*

1

Getting Started with Edge Computing on AWS

In recent years, a new operating model known as "edge computing" has emerged as a key enabler of digital transformation strategies across a variety of industries and organization types [1]. The exponential growth of data generated by an ever-increasing number of connected devices, coupled with an insatiable appetite for real-time processing and analysis, has driven new operational models in the cloud. An example of this is the **Industrial Internet of Things (IIoT)**. Also known as Industry 4.0, IIoT involves a fusion of traditional industrial processes with advanced techniques such as **Artificial Intelligence (AI)**, **Machine Learning (ML)**, and real-time data analytics.

Due to this, "edge computing" has become a buzzword. Its meaning has been diluted through overuse by vendors who want in on the opportunities presented by this emerging model of computing. This book uses the term to mean *"an architectural design that locates appropriate AWS cloud processing closer to where data is generated"*.

By disaggregating AWS services closer to the source of data generation, edge computing facilitates shorter time from data generation to decision-making, reduced system costs, and enhanced data security.

In this chapter, we're going to cover the following main topics:

- The intersection of cloud and edge computing
- The AWS edge computing strategy
- Overview of the AWS edge computing toolbox

---

1   Gartner Predicts the Future of Cloud and Edge Infrastructure

The intersection of cloud and edge computing

> **Definition of cloud computing – according to NIST**
>
> *Cloud computing is a model for enabling ubiquitous, convenient, on-demand network access to a shared pool of configurable computing resources (for example, networks, servers, storage, applications, and services) that can be rapidly provisioned and released with minimal management effort or service provider (SP) interaction.*

AWS can deliver client-server computing in the same way a utility delivers electricity to you – pay-as-you-go with no upfront investment. AWS achieves economies of scale that even the largest businesses struggle to compete with. This is true in obvious ways such as data center and server prices it can negotiate.

It is true in many non-obvious ways as well. AWS has millions of customers and an infrastructure that must be secured both within and without. AWS security operations are world-class because they have to be. When you pay for an EC2 instance, you're not just getting a virtual machine. You're also getting that world-class security operations team protecting its data physically, but also against – for example – a **Distributed Denial of Service (DDoS)** attack. Those are two things that are very costly to do in your own data center.

Your EC2 instance is also sitting right next to a panoply of additional services it can consume. Storage, networking, AI/ML, databases, analytics – the list is long[2]. It is this proximity to all that computing power and value-add services AWS provides that pushes architectures more and more toward centralization. The more things you move in-region, the greater the benefit, and thus the flywheel of migrating things into the AWS cloud.

Great – so what are the problems with centralizing all applications into an AWS region? Well, they fall into four categories: physics, economics, regulatory compliance, and good old-fashioned organizational inertia.

Physics

The speed of light is fast, but finite – a rule of thumb is to expect a 1 millisecond of latency for every 100 km of distance a signal has to travel. Keep in mind that this is rarely measured in a straight line on a map. The following diagram illustrates a common situation. In this case, a sensor is located in Oxford – just under 100 km from London:

---

2   List of AWS Services Available by Region

Figure 1.1 – Latency incurred by physical distance and indirect routing

Notice, however, that the latency from the sensor to its compute resources in the AWS region in London is 50 milliseconds, not 1. This is due to the fact that it is communicating over a mobile network using 4G/LTE. The architecture of those networks requires what is known as a backhaul to a central location where the mobile network peers with the internet. Further, due to the vagaries of how data roaming works in 4G/LTE, 40 milliseconds just to reach the internet is not unusual. Then, the packets need to travel across the internet from wherever that central location happens to be to the London region.

Take the example of a cloud-based system that manages and monitors 1 million smart cars. A sensor in a car that can detect a crash is about to occur needs a decision made quickly by a compute resource physically close to it. Waiting for a response from the remote compute in the cloud, approx. 50 milliseconds, may well be too slow. This is especially true if multiple sensors need to send in data for a multivariate decision to be made.

In such cases, physically distributing where that decision is made closer to where the sensors are is the only viable approach.

Economics

Bandwidth can be quite expensive, especially in remote locations. Sometimes, it makes sense to build aggregation points into a system to preprocess data locally before transmitting just the important parts back to a datastore in an AWS region.

Suppose an ML model is being trained on the words people say into microphones in remote locations. A digital recording of a person saying the words "bandwidth can be quite expensive" in M4A format is about 100 KB, while a text file containing those words is only 32 bytes. If all our ML model cares about is the text, we can reduce our bandwidth costs by 3,000-fold by doing that conversion in the field. If the full-sized digital recording needs to be retained, perhaps for legal reasons, it can be queued up on a storage device and physically swapped out once a month as it fills up.

Regulatory compliance

Data privacy and sovereignty requirements vary across political entities. Compliance regimes such as the EU's **General Data Protection Regulation** (**GDPR**) impose constraints upon where a given piece of data associated with an individual is physically located. For instance, a US-based company that runs its applications in the AWS cloud might decide it needs to keep records associated with EU data subjects within the physical boundaries of a country where AWS does not have a region.

Other times, national sovereignty comes into play. Countries often have special rules regarding emergency services systems, such as 911 in North America or 112 in Europe. It is not uncommon that such government-operated applications must certify that it is not possible for another country to disable them. This generally entails keeping the physical infrastructure within the borders of said country.

Inertia

Since the inception of AWS in 2006, some organizations (or individual departments within them) have taken a wait-and-see approach – maybe the cloud will just go away if they ignore it. Sometimes there are trust issues, especially when implementing hybrid cloud solutions that involve on-premise components. While this becomes less common every year, such inertia is still encountered at times.

So, given all of that, how is AWS addressing these challenges?

The AWS edge computing strategy

The combined effects of physics, economics, regulatory compliance, and organizational inertia have led to customers implementing hybrid architectures that require them to continue operating a full data center stack – even if all they want to do is host a few virtual machines.

An example of such a situation is depicted next. In this instance, the customer has built a cloud-native application making full use of the benefits afforded by AWS managed services. This application interacts with their ERP system to provide real-time insights from their manufacturing plants. However, because their ERP system must reside in a particular geography, they have retained a small physical footprint in their corporate data center. Their developers have to interact with on-premise components using different APIs than they use for cloud-based elements. Their operations team has to use a different set of tools to manage those resources, and what's worse, they have to care about the underlying substrate to a far greater degree than they do in the cloud:

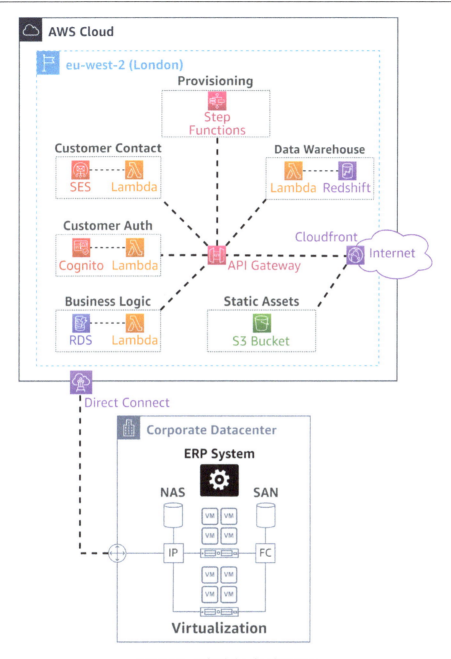

Figure 1.2 – Hybrid cloud architecture

The AWS edge computing strategy is intended to simplify this situation for customers. It does so by extending AWS' managed service offerings to physical locations outside of the core regions – to wherever customers need them.

Take the example of AWS Lambda functions. While you are likely familiar with running them in a region, they can also be run in one of AWS' 400+ points of presence for ingress/egress to its global backbone. You can take this further and run a function in your on-premise data center(s) on an AWS Outposts server or in the middle of nowhere on an embedded device such as a smart camera.

Examples of situations where this extension of AWS managed services farther from the regions can help include:

- An autonomous **Unmanned Aerial Vehicle** (**UAV**) circling over a factory scanning the perimeter for human faces and comparing them to a list of employees

- An offshore oil rig with a developing problem that local maintenance staff need to be alerted to – and even told which valve needs to be turned and at how many degrees by a digital overlay they see through their smart glasses

- An AR headset worn by a worker deep underground in a mine shaft being assisted by a geologist on the other side of the world

- A soldier observing an enemy position using a device that delivers inferences based on petabytes of cloud data to enhance that soldier's situational awareness

- In the parking lot outside of State Farm Stadium onboard a semi-trailer that acts as a miniature data center on wheels where workers are editing and producing the live video feed for Super Bowl LVII

The AWS edge computing strategy focuses on customer use cases. These are grouped into families, four of which this book will address – DDIL, MEC, immersive experiences, and IIoT.

Disconnected, denied, intermittent, or low-bandwidth

While the specific term **Disconnected, Denied, Intermittent, or Low-bandwidth** (DDIL) emerged from the US Department of Defense, it captures well the circumstances faced by a group of use cases seen across industries. It refers to edge computing in situations where network connectivity is unreliable, constrained, or completely unavailable. In such scenarios, traditional cloud-based computing approaches might not be feasible, and edge computing can play a crucial role in enabling data processing and decision-making at the source. Services such as AWS Snowball, AWS Snowcone, and AWS IoT Greengrass can help in these cases.

Disconnected

In environments where there is no network connectivity at all, devices must be able to operate independently, processing data and making decisions locally. This typically requires the implementation of efficient data storage, processing capabilities, and pre-trained ML models to allow the device to function effectively without access to the cloud. These are usually the same situations where standard data center environmental controls such as HVAC, particulate filtration, and physical security measures are lacking.

As an example, an energy company is conducting a survey for oil in a remote desert location. They are driving a special kind of truck, known as a thumper, that shakes the ground with hydraulically driven metal pads. The shock waves travel through the earth to instruments known as geophones. The data thus collected must be analyzed to produce a three-dimensional map of what is going on underground. Local ruggedized computers can be used for this first step, but the second step requires a highly trained and experienced geologist to make a judgment call about what is seen in this map – in a similar way that a radiologist is needed to properly analyze an x-ray image.

While they may be able to make voice calls or transmit data via satellite connectivity, this is both expensive and very low throughput – measured in kilobits per second. Wouldn't it be great if those determinations – inferences, in ML parlance – could be made on the spot? It would be even better if every time such a team came back into a place with better connectivity, the data they collected automatically synchronized with the cloud. This would constantly improve the experience of the system. In AI parlance, this is known as training. Such a model would necessarily be quite large, and the AWS cloud is the perfect place for this training to happen.

While most AWS services are dependent upon consistent high-speed connectivity to the AWS control plane, AWS Snow Family is different. These devices maintain their own local control- and management-plane services yet expose the same APIs and management constructs as their in-region equivalents.

Denied

In certain situations, network access can be actively denied or restricted due to security concerns, regulatory requirements, or other factors. Edge computing solutions must be able to adapt to these constraints by offering secure and compliant data processing and storage capabilities.

The classic example is forward-deployed military command posts. While in hostile territories, they may need to limit network connectivity to minimize the risk of data interception, cyber-attacks, or simply exposing their presence to the enemy. In such cases, edge computing devices can process data locally, ensuring mission-critical information remains secure and accessible. Further, pre-trained ML models can provide insights based on data gathered from thousands of disposable vibration sensors air-dropped across a hostile area – giving commanders real-time information about enemy troop movements. A combination of AWS Snow Family and AWS IoT Greengrass is ideal when such needs arise.

Intermittent/low-bandwidth

In scenarios where network connectivity is sporadic or limited, edge computing devices must be capable of managing data synchronization, processing, and storage during periods of limited connectivity, while also handling potential data conflicts and inconsistencies that might arise. AWS IoT Greengrass is perfect for a store-and-forward model such as this.

Autonomous vehicles operating in urban areas may experience intermittent connectivity due to factors such as network congestion or physical obstructions. Increasingly smaller and more powerful single-board computers can enable these vehicles to process sensor data locally and make real-time decisions, regardless of the network's reliability.

Rural healthcare facilities may have limited bandwidth, making it difficult to send large medical imaging files to the cloud for analysis. Much like the situation discussed with oil exploration, local inferences can be made in time-critical situations. Large datasets can be automatically synchronized by physical transport as time permits using members of AWS Snow Family.

Multi-access edge computing

Multi-access Edge Computing (MEC) is an innovative paradigm that brings computation and data storage capabilities closer to the edge of mobile networks, enabling real-time processing and ultra-low latency for a wide range of applications.

Mobile network operators

A **Mobile Network Operator** (MNO) is a telecommunications company that provides wireless voice and data services to its customers by owning and operating a cellular network infrastructure. MNOs are responsible for the deployment, maintenance, and management of the network, including radio access, core network components, backhaul connections, and interfaces with the internet or landline networks.

MNOs purchase radio frequency spectrum licenses from government authorities to transmit their signals and provide services such as voice calls, text messaging (SMS), multimedia messaging (MMS), and mobile internet access (3G, 4G, and 5G) to subscribers using mobile devices, such as smartphones and tablets.

Some of the largest and most well-known MNOs globally include Verizon based in North America, Vodafone based in Europe, and KDDI based in Japan. These companies play a crucial role in the telecommunications ecosystem by connecting millions of users and enabling seamless communication and data access across various geographic locations.

Emergence of 5G

Most MNOs have upgraded their networks from 4G/LTE to 5G within major metropolitan areas. However, more remote areas of the same networks are still serviced by 4G/LTE. The delay has been economic. The market for mobile devices has largely been captured. Customers want 5G but aren't willing to pay extra for it. Therefore, MNOs that do roll out 5G can steal customers from ones that don't – which ultimately means 5G rollout occurs primarily in areas of highest demand and highest competition.

This is starting to change as MNOs realize they can leverage the same upgrades they did for 5G to capture entirely new revenue streams. Historically, MNO **Central Offices** (**COs**) have been small facilities filled with purpose-built hardware for the **Radio Access Network** (**RAN**) only, with backhauls to a regional office that is an actual data center – a relic of decades-old copper-wire telco networks:

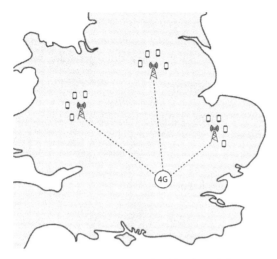

Figure 1.3 – 4G/LTE backhaul model

However, COs for an MNO moving to 5G have to be transformed into small data centers to support **Network Functions Virtualization (NFV)** and the distributed nature of 5G:

Figure 1.4 – 5G network distributed model

At the same time, more and more devices are coming online with demand for applications that require more bandwidth or do not tolerate the jitter and latency that inevitably occurs when traffic is backhauled and only then routed over the internet to a cloud provider.

COs become multitenant data centers

Enter MEC. In this model, MNOs use their existing physical footprint to provide compute and storage capacity for customer applications in the same way that cloud providers do – only much closer to the mobile devices consuming those applications. Average latency from mobile device to application is reduced to single digits, while 5G network slicing can ensure a level of jitter, security, and throughput that meets customer needs on a per-application basis:

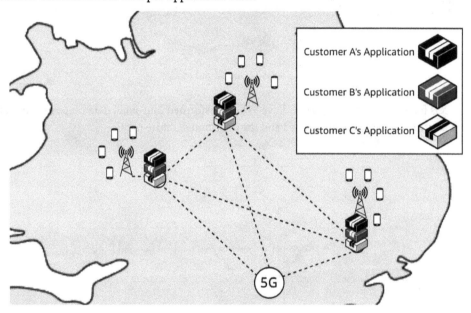

Figure 1.5 – MEC example

MEC pushes significant compute resources close to the mobile devices without requiring any change to said devices. Emerging use cases such as **Virtual Reality (VR)**, **Augmented Reality (AR)**, smart cities, and robotics benefit from lower latency. Ever-increasing demands from video streaming apps exploit MEC for pre-staging content to reduce upstream bandwidth consumption. AI/ML systems can deliver inferences closer to real-time than if bulk observations had to be shipped back to a central location.

AWS Wavelength is an end-to-end MEC service built in partnership with MNOs around the world. It allows you to activate a special type of **Availability Zone (AZ)** that can host instances or containers, and participate in a **Virtual Private Cloud (VPC)** just like a region-based AZ. Further, services such as AWS Outposts, AWS Local Zones, and AWS EKS or ECS Anywhere can be used to build your own MEC platform.

Immersive experiences

The term "immersive experiences" includes AR and VR applications, both of which present a unique set of requirements for edge computing.

AR

Not all immersive experiences involve a user wearing an expensive helmet and gloves like something out of a sci-fi movie. In fact, the majority require only a mobile device such as a phone or tablet that users are accustomed to. It's likely they privately own such devices already. AR simply means an application that overlays digital information onto your physical environment – typically using technologies such as the **Global Navigation Satellite System** (**GNSS**) and/or Wi-Fi **Access Point** (**AP**) data to triangulate your current position.

The most common example of an AR application you probably use every day is something such as Apple Maps to help you navigate. Long gone are the days of paper maps and stopping at gas stations to ask for directions. Most cars can now give you verbal directions, and some can even do things such as show a flashing red arrow on the windshield's heads-up display telling you it's time to make a right turn. There are specialized versions of such apps – for example, a maintenance worker can receive location-specific information when navigating a large industrial facility. Amazon's own fulfillment centers make heavy use of location-based AR to guide workers to the products they need to pack for a certain order.

AR technology can also enhance customer experiences in retail and e-commerce by allowing users to virtually try on products or see how items will look in their home environment. For example, customers can use their smartphones to view a piece of furniture in their living room or try on clothes virtually, helping them make more informed purchasing decisions and reducing product returns.

Educational applications are becoming AR-enhanced by overlaying digital information, such as text, images, or animations, onto physical objects or environments. For instance, students learning about anatomy can view 3D models of organs overlaid on a physical mannequin, or field technicians can receive real-time instructions and guidance when repairing complex machinery.

VR

The first place your mind probably goes when you think of the term "VR" is video gaming or entertainment, and for good reason. VR has revolutionized both industries by providing fully immersive experiences that transport users to virtual worlds. Players can engage in interactive adventures, explore fantasy landscapes, or participate in competitive sports simulations, all while feeling as though they are physically present within the game environment.

Simulation for training purposes is another area where full VR is in use today by a variety of industries. As an example, VR can provide medical professionals with a safe and controlled environment to practice surgical procedures, diagnostic techniques, or patient communication skills. By replicating real-world scenarios, VR allows trainees to build their expertise and confidence without risking patient safety or well-being.

Another use case may include engineers who are taking advantage of VR to enhance both the quality of their products and the time it takes to design them. Architects, designers, and engineers can now create and explore 3D virtual models of their projects before construction or production begins. This immersive visualization also helps non-engineering stakeholders identify potential design flaws, optimize layouts, and gain a better understanding of the finished product, ultimately saving time and resources.

Specialized requirements

Immersive experiences rely heavily on real-time user interactions, and any delay or lag can significantly degrade the user experience. AWS edge computing services can help to minimize latency by processing data closer to the source, thereby reducing the time taken to transmit data between the user's device and the data center. Services such as AWS Wavelength and AWS Local Zones bring AWS services closer to end users, ensuring low-latency processing.

AR/VR applications often involve the transmission of large volumes of data, including high-resolution graphics, audio, and sensor inputs. This requires substantial network bandwidth to ensure seamless and immersive experiences. 5G devices are capable of as much as 10 Gbps of throughput in both directions – but generally only to the first hop. AWS Wavelength allows customers to capitalize on this fact by placing instances and containers directly at that first hop. Further, services such as AWS IoT Greengrass and AWS Snow Family enable local data processing and reduce the amount of data transmitted to the cloud in the first place.

As these applications grow in popularity and complexity, the underlying infrastructure must be able to scale accordingly. AWS provides scalable edge computing solutions that can adapt to the changing demands of AR/VR workloads, ensuring optimal performance and resource utilization. Finally, immersive experience applications often involve sensitive user data, making security a critical concern. AWS edge computing solutions incorporate robust security features, such as encryption, access control, and full auditability, to protect user data and maintain compliance with industry regulations.

IIoT

IIoT refers to the integration of internet-connected sensors, devices, and machinery in industrial settings. This includes manufacturing plants, utilities, oil and gas, transportation, and agriculture. IIoT leverages advanced technologies such as big data analytics, ML, and cloud computing to collect, analyze, and transmit data from various sources, allowing for real-time decision-making, process optimization, and improved operational efficiency. By connecting industrial assets and systems, IIoT enables organizations to monitor performance, predict and prevent equipment failures, and drive overall productivity and innovation within the industry.

Message Queuing Telemetry Transport

Message Queuing Telemetry Transport (**MQTT**) is a lightweight, open source, and widely adopted messaging protocol designed for limited capacity networks, often used in **Machine-to-Machine** (**M2M**) and IoT applications where a low code footprint is required and/or network bandwidth is at a premium:

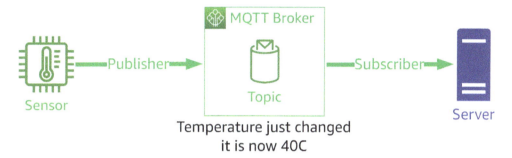

Figure 1.6 – MQTT's publish/subscribe model

MQTT operates on a publish/subscribe model, which means that devices (clients) can publish messages to a topic on a central broker, and other devices can subscribe to the broker to receive these messages. This model allows for efficient data transfer and real-time updates, without requiring a continuous connection between devices.

MQTT is renowned for its efficient use of the network, ability to operate well in unreliable environments, and ease of implementation, making it an excellent choice for edge computing, mobile applications, and other scenarios where network conditions may be challenging.

Lastly, MQTT also supports a "last will and testament" feature that allows a client to publish a message when it disconnects ungracefully. This can be useful for monitoring and managing the status of devices.

By default, MQTT uses TCP ports 1183 (unsecured) and 8883 (TLS encrypted) for reliable transport over standard IP networks or the open internet. Most IIoT devices sold these days can natively use the MQTT protocol to communicate with brokers. However, this is not always the case.

Legacy IIoT networking technologies

Legacy industrial automation systems relied on a variety of networks and protocols to facilitate communication and control among devices, machines, and systems. Examples include MODBUS, PROFINET, EtherCAT, and Fieldbus. These networks were typically closed systems – islands unto themselves. They were also developed at a time when such devices were limited in number and directly connected via serial cable:

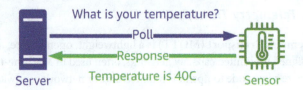

Figure 1.7 – Legacy poll/response IIoT protocols

This is why they typically use a poll/response protocol architecture. Unlike the publish/subscribe and report-by-exception techniques MQTT uses, these protocols require a server to periodically query each device. Those devices, in turn, respond with the requested data.

Poll/response protocols have a number of disadvantages compared to MQTT:

- **Efficiency**: Legacy poll/response protocols require the client to continually ask (poll) the server for data. This can lead to excessive network traffic and wastage of bandwidth, especially when dealing with a large number of devices. MQTT, on the other hand, only publishes updates by exception. For example, an MQTT-based temperature sensor will only publish a new reading if the temperature changes – in a poll/response model, there could be hundreds of duplicate readings sent over the network.

- **Scalability**: Legacy protocols struggle to scale such that they can handle the high number of devices and data points typically found in today's IIoT environments[3]. MQTT was designed with scalability in mind and can easily handle communication between a large number of devices.

- **Real-time updates**: Legacy poll/response protocols may not provide real-time updates as they work on the request/response model. With MQTT, updates can be pushed immediately when the data changes, enabling near real-time communication.

- **Network resilience**: Poll/response protocols offer no guarantee that the data was received. MQTT is designed to work over unreliable networks, and with features such as **Quality-of-Service (QoS)** levels and "last will and testament," it ensures data delivery even in challenging network conditions.

- **Security**: Many legacy protocols were not designed with security in mind and do not include built-in security features such as encryption or outbound-only connection initiation, making them vulnerable to cyber-attacks. MQTT uses TCP/IP and supports modern security mechanisms such as SSL/TLS for secure data transmission. MQTT also does not require inbound TCP/UDP ports on clients to be open, as connections are always initiated from the client to the broker.

- **State awareness**: With poll/response, the server does not know the state of the client, while the MQTT broker keeps track of client sessions and their subscriptions, improving the overall awareness of the system's state.

---

3   How MQTT is Helping Support the Digital Transformation

- **Flexibility**: Legacy protocols may not support diverse data types and structures, limiting flexibility. MQTT, on the other hand, is data-agnostic and can support various data types and complex nested structures, making it more versatile for modern IIoT applications.

Despite these advantages, the older devices do continue to function and often represent a significant capital investment. It is unusual for IIoT deployments to call for a total replacement of those past capital expenses.

Supervisory Control and Data Acquisition

Supervisory Control and Data Acquisition (SCADA) systems are used in industrial settings to monitor and control processes and infrastructure in various industries – including manufacturing, utilities, oil and gas, water management, and transportation. SCADA systems enable real-time data collection and processing from remote equipment, allowing operators to supervise and manage industrial processes from a centralized location:

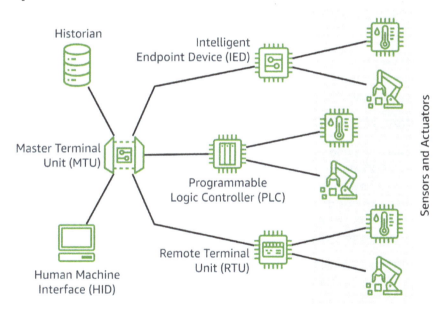

Figure 1.8 – Typical SCADA system architecture

A typical SCADA system consists of several components, including the following:

- **Remote Terminal Units (RTUs)**, **Intelligent Electronic Devices (IEDs)**, or **Programmable Logic Controllers (PLCs)**

  These devices are responsible for collecting data from sensors and controlling equipment at remote sites. They communicate with the central SCADA system to transmit data and receive control commands.

- **Sensors and actuators**

  Sensors measure various process parameters, such as temperature, pressure, and flow rate, while actuators perform control actions, such as opening or closing valves and adjusting motor speeds. These devices interface with the RTUs or PLCs to provide data and receive control signals.

- **Human-machine interface (HMI)**

  An HMI is a graphical interface that allows operators to interact with the SCADA system, visualizing process data, and issuing control commands. HMIs typically display real-time data in the form of charts, graphs, and schematics, enabling operators to monitor system status and make informed decisions.

- **Historians**

  A historian is a system that holds an archive of past data samples taken from PLCs, RTUs, sensors, actuators, and the SCADA system itself. These are similar to **Time Series Databases (TSDBs)** such as CrateDB or TimescaleDB.

- **SCADA server**

  The SCADA server hosts the software that manages data collection, processing, and storage. It also handles alarm management, event logging, and reporting functions. The software enables operators to configure the system, define control logic, and analyze historical data.

AWS integration with SCADA systems

While there has been an explosion in the number of inexpensive sensors that natively speak IoT protocols such as MQTT, doing IIoT with AWS does not entail a complete replacement of existing systems. Services such as AWS IoT SiteWise Edge help customers interface with these protocols to merge existing data sources with new ones. In fact, the analysis of historical data from these systems helps customers realize value more quickly than starting from ground zero. AWS IoT TwinMaker can, in turn, consume this normalized data to produce digital twins of industrial operations.

When store-and-forward capabilities are desirable, or ML inferences need to be made locally, AWS IoT Greengrass is an ideal solution. It can run on your own hardware, or hardware provided by AWS in the form of AWS Outposts and AWS Snow Family.

As use cases go, DDIL, MEC, AR/VR, and IIoT all seem rather distinct from one another. How does AWS go about taking a unified approach to helping customers with all four? This is what we'll cover in the next section.

Overview of the AWS edge computing toolbox

The AWS edge computing strategy aims to provide a comprehensive suite of services and solutions that enable businesses to harness the power of edge computing, addressing the challenges of data processing, latency, security, and scalability. By bringing AWS services and resources closer to end users and devices, this strategy allows organizations to optimize their applications and infrastructure for improved performance, efficiency, and user experience.

Key components of the AWS edge computing strategy include the following topics.

Localized AWS infrastructure and services

AWS offers solutions such as AWS Outposts, AWS Local Zones, and AWS Wavelength to extend the AWS cloud infrastructure to on-premises environments, local data centers, and even mobile networks. These localized infrastructure solutions enable lower latency, reduced data transfer costs, and better compliance with data sovereignty regulations.

AWS Snow Family facilitates edge computing by enabling local data processing, storage, inferences, and analytics without a reliable connection back to the AWS control plane.

Finally, AWS IoT Greengrass can be visualized as a platform for running tiny versions of region-based AWS services you are already familiar with on devices with single-board computers – think Raspberry Pi or Arduino. This includes Amazon SageMaker, AWS Lambda, AWS Kinesis Data Firehose, AWS Kinesis Video Streams, Amazon SNS, AWS Secrets Manager, Amazon CloudWatch, and AWS Systems Manager.

Developer tools and resources

AWS offers a rich ecosystem of developer tools, SDKs, and resources to simplify the development, deployment, and management of edge computing applications. These tools and resources help developers build, test, and monitor applications for edge devices and environments, ensuring seamless integration with AWS cloud services.

Security and compliance

AWS places a strong emphasis on security and compliance, providing robust encryption, access control, and monitoring features for edge computing solutions. This allows organizations to safeguard their data, infrastructure, and applications, while also adhering to industry-specific regulatory requirements.

Consistent experience

The AWS edge computing strategy seamlessly integrates with the broader AWS ecosystem, ensuring that organizations can take advantage of a wide range of AWS services, from compute and storage to analytics and ML, to support their edge computing needs.

Architectural guidance

When architecting a solution, one should always start from a project's requirements, taking into account situational constraints, risk tolerance, and any assumptions that have been made. It is not possible to architect for all scenarios in which edge computing is useful using a single approach. It solves different problems in different ways depending on the use case.

The AWS Well-Architected Framework addresses patterns and anti-patterns with such an approach in mind. It is an AWS-specific version of more general enterprise architecture frameworks that have existed for decades such as Zachman or TOGAF – if you are familiar with those, you will see their DNA embedded in it.

If you aren't familiar, don't worry – just know that the architectural guidance presented both in this book and by the AWS Well-Architected Framework generally is built upon decades of experience in what sort of things need to be considered for the proper function of any system.

Summary

In this chapter, we discussed the opportunities and challenges that edge computing presents. This includes those imposed by the laws of physics, the laws of economics, and the law of the land. We covered the AWS edge computing strategy, which is to extend its infrastructure and services closer to end users, allowing developers and operations staff to have a consistent experience via common APIs and management constructs.

We covered existing IIoT systems and protocols and introduced you to how AWS can integrate with them. We explored how MEC with 5G networks offers a unique opportunity for both MNOs and application developers to extend cloud applications closer to users on mobile devices. We also discussed the challenges associated with DDIL use cases in edge computing, as well as the transformative impact of immersive experiences and given example use cases you may not have considered before.

Finally, we briefly introduced you to AWS Local Zones, AWS Outposts, AWS Snow Family, AWS Wavelength, and AWS IoT Greengrass and placed them in the context of the three primary edge computing use case families.

In the next chapter, we will explore issues associated with networking and security for near-edge computing scenarios.

2
Understanding Network and Security for Near-Edge Computing

There are two types of edge computing where the cloud is concerned – near the cloud and far from the cloud. Near-edge networking assumes reliable high-speed access and is probably what you are familiar with. Servers connecting from a data center over **Multiprotocol Label Switching** (**MPLS**), a laptop in a home worker's apartment with a cable modem, or a mobile device connecting via 5G or Wi-Fi are all examples of near-edge networking.

In this chapter, we're going to cover the following main topics:

- Understanding internet challenges
- Using a private wide-area network
- Optimizing ingress with global server load balancing
- Steering traffic at the network layer with IP Anycast
- Addressing TCP issues with HTTP/3 and QUIC
- Securing networks at the near edge

Understanding internet challenges

Any device that wants to use resources in the cloud must have a connection to the internet. This is sometimes done over country-specific sovereign networks for sensitive government-related workloads, but most of the time, the connections are done over the public internet. This is far more cost-effective than building a bespoke global network of your own, but the downside is that you don't get to engineer the internet.

On the internet, once your traffic goes to your ISP, you trust them and all of the other systems between them and your destination to get your traffic where it is going. This is by design. It works out fine – most of the time. Your traffic may take a less-than-ideal path along the way, increasing its latency. Or perhaps, while crossing a dozen devices to get where it's going, one packet out of a thousand is lost and has to be retransmitted.

The impact of latency and packet loss

One in a thousand (0.1%) packet loss might sound like a small problem. Any amount of packet loss limits the effective throughput that can be achieved for a given flow across the internet – and a 0.1% loss is considered quite high. This problem is further magnified when the distance between the client and the server is higher:

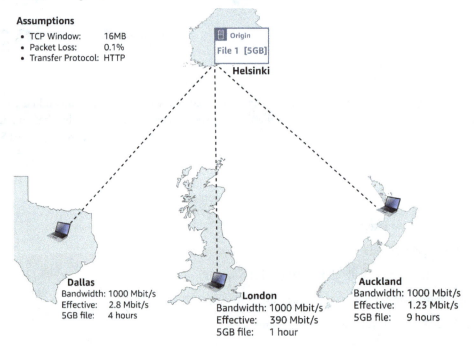

Figure 2.1 – Impact of latency and packet loss on HTTP transfers

Figure 2.1 shows three examples of clients experiencing 0.1% packet loss in different parts of the world. Even though each client has a 1 Gbit/s fiber connection to their ISP, the effective throughput that can be achieved varies widely due to different amounts of latency compounding the problem to different degrees. To understand how these values were arrived at, we need to review some of the factors that determine the effective throughput of any given TCP connection.

> **Round-trip time (RTT)**
>
> Also known as latency, this represents the amount of time in milliseconds that a packet takes to travel from sender to receiver and back again.

Assuming the path between the sender and the receiver is a straight line, there is no way to reduce RTT because it is limited by the speed of light. The speed of light is fast, but finite – for every 100 km (60 miles) of distance a signal travels, 1 millisecond is added to its RTT.

> **Maximum segment size (MSS)**
>
> MSS is the maximum number of bytes that may be contained in the payload section of a TCP segment.

To calculate the appropriate MSS value, you must take the standard **Maximum Transmission Unit** (**MTU**) of 1,500[1] and subtract any overhead involved.

Here is an example

Layer 4: TCP header of 32 bytes[2]

Layer 3: IP header of 20 bytes

Layer 3: 78 bytes overhead if an IPSEC VPN tunnel is being used

In this example, the MSS value would be 1,500 – 32 -20 -73 = 1,375 bytes

The MSS value is set in the receiver's operating system. That value is announced by the receiver during the TCP three-way handshake, telling the sender it is the largest payload it can accept. This can be restricted to a desired value by a stateful device in the middle, such as a firewall or load balancer, via MSS clamping.

---

1   The path MTU over the internet is almost always 1,500 bytes.

2   TCP headers are 20 bytes natively, but these days, it is a safe assumption that TCP timestamps are in use, which raise the value to 32 bytes.

Mathis equation

The maximum throughput that can be achieved by a TCP connection can be calculated as follows:

$$t = \frac{m}{r} * \frac{1}{\sqrt{p}}$$

Here, we have the following:

- t is the effective throughput in bits per second
- m is the MSS value in bits
- r is the RTT value in milliseconds
- p is the percent packet loss as a decimal number

Figure 2.2 below illustrates how the interaction of RTT and packet loss affects the effective throughput of a connection. The lighter line shows a loss of one packet in a million, while the darker line shows a loss of one in a hundred thousand.

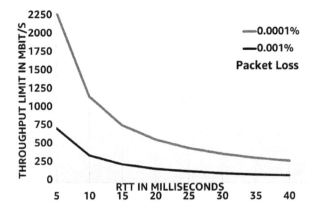

Figure 2.2 – Effective throughput estimated by the Mathis equation

Figure 2.2 illustrates how the interaction of RTT and packet loss affects the effective throughput of a connection. The lighter line shows a loss of one packet in a million, while the darker line shows a loss of one in a hundred thousand. Remember, under these conditions, it does not matter how fast your internet connection is – these are hard limits. As you can see, it doesn't take much packet loss to severely curtail an end user's experience.

Causes of packet loss on the internet

The primary cause of packet loss on the internet is congestion or throttling at a peering point between two **Autonomous Systems (ASs)**. ASs are typically operated by a single large organization such as an ISP, a large technology company, a cloud service provider, a university, or a government agency. Every time traffic crosses the boundary between two ASs, the odds of the overall flow experiencing packet loss somewhere along the way increase:

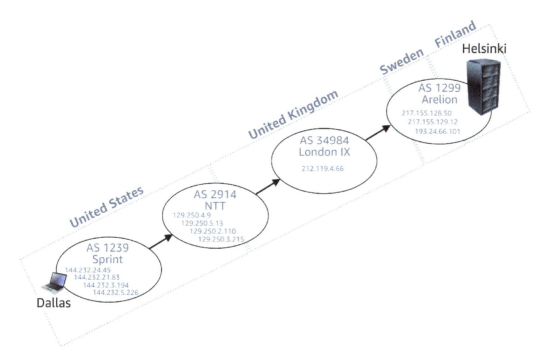

Figure 2.3 – Example of AS traversal on the internet

Figure 2.3 elaborates upon the path taken by our hypothetical client in Dallas. From this client to the server, there are a total of 12 routing hops. Not all of the hops are equivalent, though. The first four hops are within Sprint's network, while the fifth hop is on AS 2914, which is operated by NTT. That junction between providers is what is known as a peering point. Routing hops that traverse peering points are much more likely to introduce packet loss due to congestion than intra-AS routing hops.

TCP receive window (RWIN)

Even on a perfectly clean network, there's another thing that can artificially limit throughput when the RTT gets high. In TCP, part of the ongoing conversation between the sender and the receiver is something called the *TCP receive window* (RWIN). It is a value that can range from zero to 1,073,725,440 bytes, or about 1 GB[3].

It represents the amount of total amount of unacknowledged data a sender may have in flight before it must stop and wait for the receiver to send an ACK message for one or more of the already sent TCP segments. This is why it is also called the *receive buffer*.

---

3   Technically, the maximum RWIN value is 65,535 bytes, but it is multiplied by the window scaling value, which can be up to 16,384. For example, an RWIN of 65,535 bytes with a window scaling value of 2500 would result in an effective window size of 163,837,500 bytes or ~163MB.

On connections with a high RTT, this can lead to situations where the sender has to stop and wait so often that the effective throughput is noticeably impacted.

While packet loss due to congestion or throttling is possible across the transit links within a provider's network, for a host of reasons, it is most often observed at the peering points:

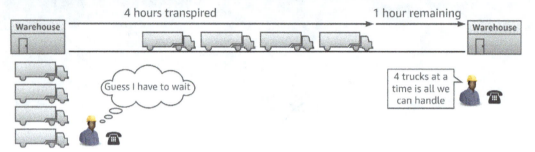

Figure 2.4 – TCP receive window and latency

Why this leads to problems with long RTTs isn't always intuitive. Therefore, we've drawn up an analogy in *Figure 2.4*. In this scenario, the sender can see that there is room on the road for more trucks, but they aren't allowed to send more until they receive a phone call from the receiving warehouse saying it is okay to do so. There are multiple reasons this could be the case.

Here are some possibilities for any given truck:

- It arrived, but it hasn't been unloaded yet as the dock is overwhelmed

- It's just a long trip and we need to be patient – they are still on the way

- It crashed somewhere in the middle or was hijacked and is lost forever

Ideally, the receiving warehouse will call at some point and say they've successfully unloaded the cargo (TCP ACK) and what they unloaded matches the manifest (TCP checksum match). In this case, the sender will cross that one off the list and send the next truck waiting to go.

If enough time goes by without that phone call, the sender will decide a truck has been lost (TCP timeout) and send a replacement (TCP retransmit). If this happens enough times, the sender might decide that the problem is a traffic jam in the middle that they might be contributing to. Therefore, they need to start waiting longer and longer (TCP retransmission timeout) until the lost truck issue stops happening for a while (TCP congestion control using exponential backoff):

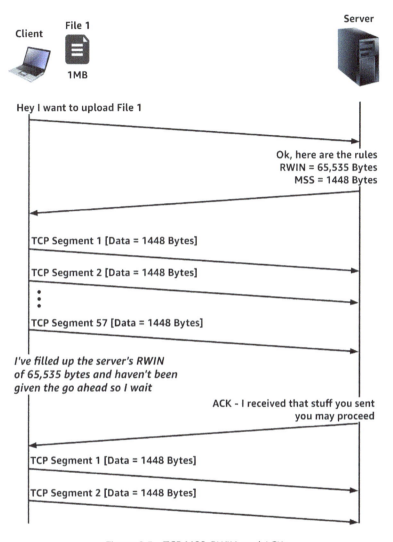

Figure 2.5 – TCP MSS, RWIN, and ACKs

Because the congestion and latency situation is different for every connection on the internet; modern operating systems usually do not have a set RWIN size in their TCP stack. Rather, they dynamically ramp up or down throughout the connection.

For example, if the receiver's buffer keeps filling up due to memory issues, it may set a lower RWIN to slow the sender down. Alternatively, the receiver may start with a low RWIN at first and keep increasing it until retransmissions occur, which likely indicates congestion in the middle. At this point, it will back off a little to find the sweet spot. How much of a problem a retransmission is depends on the size of the window – having to resend an MB of data is a different story than having to resend a GB. That's why finding and maintaining an optimal window size takes many things into account.

TCP RWIN formula

Here's the TCP RWIN formula:

$$t \leq \frac{w}{r}$$

Here, we have the following:

- t is the throughput in megabits per second
- w is the RWIN in kilobits
- r is the RTT value in milliseconds

When you're trying to work out the effective throughput of any connection, both calculations must be performed. First, apply both the Mathis equation for packet loss, then apply the TCP RWIN formula. Whichever result is lower is the maximum throughput that connection will attain, regardless of the available bandwidth:

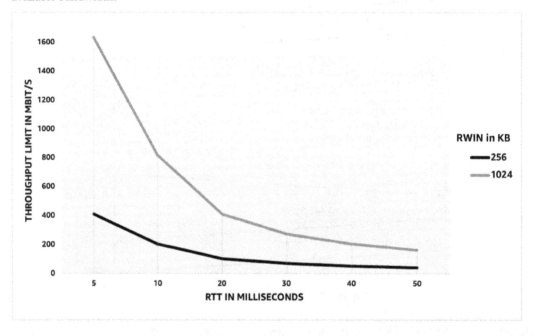

Figure 2.6 – Effective throughput estimated by the RWIN formula

Figure 2.6 demonstrates how the RWIN size conspires with higher RTTs to severely curtail maximum throughput. The lighter line on top shows an RWIN of 1,024 KB (or 1 MB), while the darker line represents an RWIN of 256 KB. TCP RWIN imposes hard limits on throughput in the same way that packet loss does.

User Datagram Protocol (UDP)

UDP is a common alternative to TCP that also operates at Layer 4 of the OSI model. Unlike TCP, UDP does not establish a connection – there is no three-way handshake. This means RTT and packet loss do not artificially limit the throughput a connection can achieve. One UDP stream is perfectly capable of filling the entire end-to-end bandwidth available to it.

The downside is that there is no error detection, no congestion control, no guarantee of delivery or ordering, and duplicate datagrams are not detected by it. This means the application needs to handle those things itself. Those are nontrivial things to implement per application. This is why applications that require reliable data transfer tend to rely on TCP to handle that for them.

However, there are applications for which reliable transmission doesn't matter as much. **Voice-over-IP** (**VOIP**) applications such as Skype, WhatsApp, or FaceTime will just move on if some datagrams are lost. Online games such as Fortnite or Call of Duty use UDP because, again, once a datagram is lost, the real-time situation has moved on anyway. Applications such as YouTube, Netflix, and Hulu use UDP for their video streaming for that same reason.

Using a private wide-area network (WAN)

When the open internet proves too unreliable in terms of latency, jitter, packet loss, or path convergence, the historical answer has been for companies to build a private WAN using MPLS.

Multiprotocol Label Switching (MPLS)

Whereas IP routing operates at Layer 3 of the OSI model, MPLS operates below that (often called Layer 2.5). Data is forwarded based on labels along predetermined paths, which allows MPLS to offer far more reliable packet delivery than IP routing over the internet can.

The trouble with MPLS is bandwidth cost. Because many organizations do not have the capital required to deploy their own global MPLS network, paying an ISP or telco for a slice of theirs is common. However, this is quite expensive. For reference, 1 Gbps of business-class internet access from an ISP in New York City might cost $500/month. 1 Gbps of MPLS service in the same city could easily cost $100,000/month. This is why MPLS tends to be purchased in much smaller increments and used only for mission-critical traffic.

Software-defined networking (SDN)

Manually provisioning a new application on a large enterprise network requires several steps. Each one may well be handled by a different specialist. Of course, every person involved needs at least a couple of days to respond. Here is an example of how provisioning workflows are born:

1. A ticket is sent to the infrastructure team requesting a VLAN for new app servers (48-hour SLA).
2. A ticket is sent to the security team requesting firewall rules for dependencies (48-hour SLA).

3. A ticket is sent to the WAN team requesting MPLS routing to the remote database (48-hour SLA).

4. A ticket is sent to the platform team requesting a server pool on F5 (48-hour SLA).

5. A ticket is sent to the network team requesting an SNAT pointing to the F5 (48-hour SLA).

This equates to 10 working days or two calendar weeks. Factor in someone going on vacation or being out sick, or some approval being needed somewhere in the middle, and a full month is not uncommon – just to get the network laid down for this new application.

Automation is needed, but each step requires a specialist to touch an expensive, vertically-scaled, and highly complex piece of equipment. It is unlikely that all parties are going to agree to let you send API commands to shared mission-critical hardware such as their core switch when one wrong move can take the entire data center down. Even if they all did, the diversity of hardware vendors involved means you would likely end up maintaining a provisioning system built of duct tape and baling wire.

What many enterprises, and all cloud service providers, do to address this is use an overlay transport of some type. It essentially operates as a VPN mesh (though it is not always encrypted) that abstracts the physical layer, which becomes known as the underlay. Its only job now is to move packets from A to B. VXLAN, Geneve, IPSEC, and other such protocols can be used for this purpose.

Now, entirely virtual versions of VLANs, firewalls, load balancers, switches, routers, and so on can be deployed to perform the same functions that hardware appliances used to. Only now, because they are virtual, you can have one for every application.

Because the virtual network constructs are software, automated provisioning is straightforward – if for no other reason than because the blast radius of any problems during provisioning is contained to only that one application. Even better, the entire end-to-end configuration of what used to involve five tickets and three to four weeks can now be deployed from a template with a single click.

Finally, because the network constructs are entirely virtual, it is possible to eliminate the need to hairpin out to an east-west firewall or router just to come right back and communicate with a different virtual machine on the same hypervisor because it happens to be connected to a different VLAN. Now, that sort of logic can be distributed down to the hypervisors themselves, which can offload significant amounts of unnecessary east-west traffic in the data center.

Software-defined WAN (SD-WAN)

SD-WAN is the application of SDN principles to networks outside your data center or span of control. The problems that SDN solves within your data center are even larger challenges on the internet, where you definitely won't be given access to modify equipment you don't own.

The general idea is to implement a logical overlay between one site and another on the internet. This is very similar to how a standard VPN works. The difference is that SD-WAN tunnels can span multiple networks, both public and private:

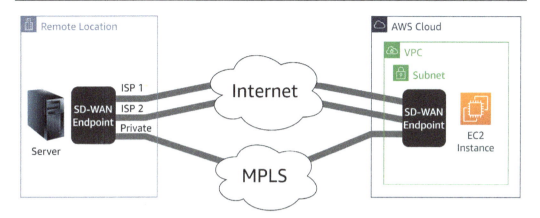

Figure 2.7 – SD-WAN spanning public and private networks

Figure 2.7 shows an example of this. From the perspective of the server on the left, there is a single Layer 3 hop to reach the EC2 instance on the right. However, the underlying situation is more complex. Connections to the internet from two different ISPs and a private MPLS fabric are all used simultaneously.

The following section describes some reasons you might do this.

Improved performance

SD-WAN allows intelligent path selection and dynamically routing traffic based on application requirements and network conditions. This enables the efficient use of available bandwidth, ensuring optimal performance for critical applications, and reducing latency and packet loss.

Cost savings

By using SD-WAN, organizations can reduce their reliance on expensive, dedicated MPLS circuits for WAN connectivity. Instead, they can leverage a combination of lower-cost broadband, fiber, or cellular connections to achieve similar performance levels at a reduced cost. Alternatively, if MPLS is still used, packets can be selectively routed over it based on, say, how mission-critical the traffic is, thereby lowering the bandwidth needed for MPLS.

Simplified management

SD-WAN centralizes the management and control of the network, providing a unified interface for administrators to monitor, configure, and troubleshoot the entire WAN infrastructure. This simplification reduces the need for manual intervention and makes it easier to manage complex networks.

Enhanced security

SD-WAN provides built-in security features, such as end-to-end encryption and segmentation, to protect sensitive data as it traverses the network.

Optimizing ingress with global server load balancing (GSLB)

Consider the situation shown in *Figure 2.8*:

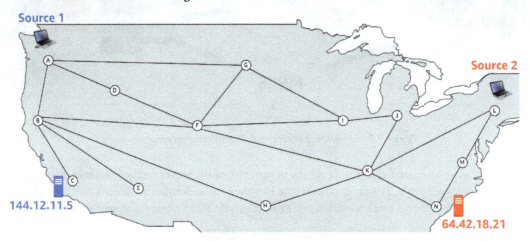

Figure 2.8 – A geographically distributed application

This application has a server on the west coast of the US, and another one on the east coast. How do we make sure Source 1 in Seattle goes to Server 1 in Los Angeles while Source 2 in NYC goes to Server 2 in Atlanta? The simplest answer would be to tell Customer 1 to go to server1.myapp.io and tell Customer 2 to go to server2.myapp.io. That might work for a handful of customers:

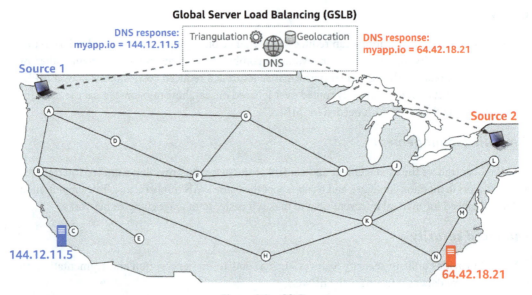

Figure 2.9 – GSLB

Realistically, though, we probably want a single address of myapp.io. How do we make it so both customers automatically get routed to the server that is geographically closest to them? Historically, there have been two main approaches: GSLB and IP Anycast.

GSLB is just DNS with some extra intelligence added. *Figure 2.9* shows how it works. The GSLB servers are set as the **Name Server** (**NS**) for myapp.io with your DNS registrar. When DNS requests come in, it will use one or more techniques to figure out which IP address to respond with.

Internet Control Message Protocol (ICMP) triangulation

ICMP, informally known as "pinging," is an error-reporting protocol that's used by network devices and operating systems to figure out whether a given IP address is reachable, whether it is experiencing packet loss, or how far away it is in terms of latency.

With ICMP triangulation, the IP that issued the DNS request is simultaneously pinged from the data centers where each server is located. Whichever ping response shows the lowest RTT is deemed the closest. This method results in the best performance, but it doesn't care about where the client truly is physically located, just that it has the least latency.

IP geolocation database

Several companies maintain databases that map IP ranges to physical locations and update this data at regular intervals. GSLB appliances typically pay for a subscription to these feeds. But where do those companies get their data? When an ISP obtains public IP space or an ASN, it ultimately does so via the **Internet Assigned Numbers Authority** (**IANA**), and WHOIS queries to IANA for the owner of the IP prefix and ASN in question are a good source. Another data source emerges from the analysis of reverse DNS crawls. Most ISPs name their routers things such as `r1.dfw.tx.isp.com`, which gives you a pretty good hint that a client connected to that router is somewhere near Dallas, TX, USA.

If you are using an IP geolocation database because you need to make sure customers go to a specific server that is physically located in a certain country for compliance reasons, you probably don't care as much about minimizing latency. VPNs pose a problem for these situations as they mask the true IP of the device with that of a VPN gateway that could be located on the other side of the world. Proxy detection databases are another offering that GSLB users can subscribe to so that they know how to refuse service to clients coming from an IP known to be owned by a VPN service provider or similar.

Custom rules

Most GSLB solutions can incorporate some form of logic above and beyond the other methods that would be specific to a given deployment. For example, if Server 1 is under heavy load, we might decide to send Customer 1 to Server 2 even though the RTT is higher. We might also parse the URL or IP header information and make a decision based on that – for instance, we might pin a remote worker to a certain home server regardless of where they travel in the world for compliance reasons.

Of course, none of these methods are perfect. If someone wants to set up a VPN server in another country and move it to a new IP if it ever winds up on a VPN server list, it is next to impossible for GSLB to detect this.

Physical or virtual appliances that are capable of doing GSLB are known as **Application Delivery Controllers (ADCs)** or **Global Traffic Managers (GTMs)**. While it is possible to build a functioning GSLB solution yourself with open source software, it is complex to both deploy and operate. For something so mission-critical, this is a risk most enterprises are not willing to take. Therefore, fully integrated solutions from vendors such as F5, Citrix, A10, Infoblox, and the like are the most common choice for self-managed GSLBs, despite their significant upfront cost.

Steering traffic at the network layer with IP Anycast

IP Anycast is another common approach for global traffic distribution. *Figure 2.13* depicts an example of how it works:

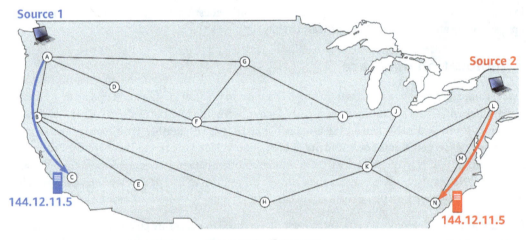

Figure 2.10 – IP Anycast

Both servers have a public IP address of 144.12.11.5. Customers are automatically routed to the closest one. This makes DNS simple; you just need a single A record for myapp.io pointing to that IP.

You may be asking yourself: wait, I thought public IP addresses had to be unique on the internet? That is normally true. However, if one has access to their provider-independent IP space from IANA, some tricks can be utilized.

Recall the situation in *Figure 2.3*, which shows the peering that happens between different ASs on the internet. When that peering happens, one AS tells another AS about the IP prefixes inside of it. It also passes along BGP community information, which is a series of variables that can override the normal mechanism BGP uses to figure out the best path to get to a destination within that AS. This information can be manipulated in such a way as to make all other ASs think whatever we want them to think about how to get to an IP we control.

However, if your AS peers with multiple ISPs, you will need to get both of them to cooperate concerning how your IP prefixes are advertised. This compounds the already difficult task of obtaining your own public IP space from IANA in the first place. Thus, IP Anycast is typically the province of the ISPs themselves. Most organizations lease individual IP Anycast addresses from their ISPs:

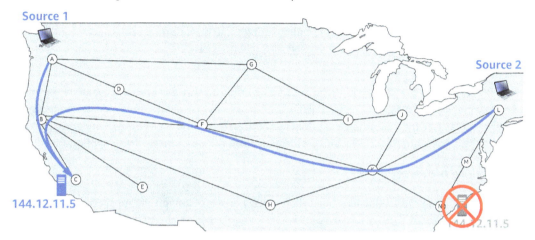

Figure 2.11 – IP Anycast path convergence

One of the reasons you might choose IP Anycast over GSLB is because it has a shorter path convergence time. With GSLB, if one of the data centers goes offline, the GSLB server needs to realize this and then start handing out a different IP – but due to the way DNS caching works on devices and at their ISPs, the timeout for this to happen can be lengthy. Unless you can enforce low **Time-To-Live** (TTL) values on the cache of all intermediate DNS servers *and* keep device cache times short (an unlikely prospect), they will be re-routed to the other data center more quickly with IP Anycast.

One of the main downsides to IP Anycast is that the routing decisions are only made in the network, which is not typically something an application owner has an understanding of or the access to modify if they did. It is more straightforward for a developer to specify the behavior they want with GSLB, and more parameters can be taken into consideration.

Addressing TCP issues with HTTP/3 and QUIC

HTTP/3 – Hypertext Transfer Protocol version 3

This is the latest revision of the HTTP protocol and is widely used for communication between web browsers and servers. It is based on **Quick UDP Internet Connections** (QUIC), a transport protocol developed by Google. QUIC is designed to provide a secure and efficient transport layer protocol over the internet.

Upsides of HTTP/3 and QUIC

QUIC, being based on UDP, doesn't suffer from throughput limitations due to latency or packet loss seen with TCP-based protocols such as HTTP/2. This is its primary benefit as it relates to edge computing – you no longer need to do a bunch of calculations and mitigations to accommodate these factors.

However, there are many other benefits to implementing these newer protocols.

Connection setup latency

In HTTP/2, establishing a connection requires a series of round trips between the client and server, leading to increased latency. QUIC, being built on UDP instead of TCP, significantly reduces connection establishment latency by combining the initial connection setup and encryption handshake into a single step. This helps improve the overall performance and responsiveness of web applications.

Head-of-line blocking

HTTP/2 suffers from head-of-line blocking, where a delay or loss of a single packet affects the delivery of subsequent packets. QUIC resolves this issue by using packet-level parallelism. Each packet in QUIC is treated as an independent unit, enabling concurrent delivery and reducing the impact of packet loss or delay on other packets. This improves overall throughput and minimizes the effect of network congestion.

Packet loss and recovery

TCP, the underlying transport protocol of HTTP/2, relies on congestion control mechanisms that can be overly cautious and slow to recover from packet loss. QUIC includes congestion control algorithms that are specifically designed for the characteristics of modern networks. It uses forward error correction and retransmission mechanisms to recover lost packets more efficiently, resulting in improved reliability and reduced latency.

Security

QUIC integrates encryption by default, providing secure communication between clients and servers. Unlike HTTP/2, which relies on additional protocols such as TLS to establish secure connections, QUIC ensures end-to-end encryption without the need for separate encryption layers. This enhances security and privacy for data transmission over the internet.

Network traversal

QUIC operates at the transport layer and is designed to work seamlessly with modern network infrastructures, including those with **Network Address Translation** (**NAT**) and firewalls. It encapsulates the QUIC packets within UDP, making it easier to traverse network boundaries without requiring complex configuration changes. This facilitates faster deployment and adoption of the protocol in various network environments.

Multiplexing and stream management

HTTP/2 introduced multiplexing, allowing multiple streams to be sent concurrently over a single connection. However, managing streams and their dependencies can become complex, leading to suboptimal performance. QUIC improves upon this by providing more efficient multiplexing and stream management. It allows for independent flow control and enables better prioritization of streams, ensuring optimal utilization of available network resources:

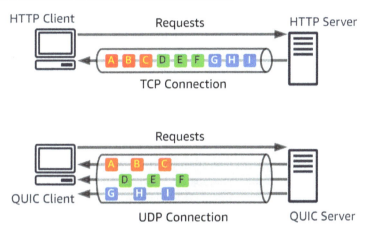

Figure 2.12 – Comparison of HTTP/1.1 and HTTP/2 with QUIC

By addressing these issues, HTTP/3 and QUIC aim to enhance the performance, security, and reliability of web applications. The protocol's ability to reduce latency, mitigate the impact of packet loss, provide built-in encryption, and offer efficient network traversal makes it well-suited for modern web communications, especially in scenarios with unreliable or congested networks. The widespread adoption of HTTP/3 and QUIC is expected to contribute to faster, more secure, and more efficient communications on the internet.

Downsides of HTTP/3 and QUIC

While HTTP/3 and QUIC offer numerous benefits for web communication, there are also some drawbacks and challenges associated with their increased use on the internet. Let's look at a few considerations.

Limited adoption

HTTP/3 and QUIC are relatively new protocols, and their adoption is still in progress. As a result, not all web servers, clients, and network infrastructure fully support these protocols. This limited compatibility may pose challenges when you're trying to establish connections with systems that do not support HTTP/3 or QUIC yet. So, when implementing HTTP/3, it is important to include mechanisms that redirect incompatible clients to HTTP/1.1 or HTTP/2 versions of the service in question.

Migration complexity

Migrating from older protocols, such as HTTP/1.1 or HTTP/2, to HTTP/3 and QUIC can be complex and time-consuming. It requires updating server software, client libraries, and network infrastructure. This migration process may involve compatibility issues and necessitate careful planning and testing to ensure a smooth transition.

Increased resource consumption

While HTTP/3 and QUIC are designed to improve performance, they may also lead to increased resource consumption compared to older protocols. The additional encryption and packet-level parallelism introduce some overhead, requiring more processing power and bandwidth. This could impact the performance of resource-constrained devices or networks with limited resources.

Network congestion

The use of UDP as the underlying transport protocol for QUIC can potentially lead to network congestion issues. UDP packets bypass congestion control mechanisms used by TCP, leading to increased congestion in networks that aren't prepared. However, QUIC includes its own congestion control mechanisms, which attempt to mitigate this concern. Monitoring and fine-tuning these mechanisms is crucial to maintaining a fair and efficient network.

Security considerations

The inbuilt encryption makes it more difficult to inspect network traffic for security purposes, such as intrusion detection or deep packet inspection. Network administrators and security professionals may need to adapt their monitoring and security practices to accommodate the encrypted nature of HTTP/3 and QUIC.

Increased complexity for troubleshooting

The layered nature of these protocols and their interactions with various network components make diagnosing and resolving problems more challenging. Network administrators and developers may need to acquire new skills and tools to effectively troubleshoot issues specific to HTTP/3 and QUIC.

Current status

Download Wireshark and do a packet capture while browsing your favorite websites. One thing will be apparent – almost any video-streaming service, such as NetFlix, YouTube, Hulu, or Amazon Prime, makes heavy use of QUIC. At the time of writing, content delivery networks, and the media industry in particular, have been leading the way in adoption. However, this hasn't been the case with enterprise applications in general.

As the protocols continue to evolve and gain broader support, their drawbacks are expected to diminish over time. Many of the challenges are being actively addressed, although, like any sea-change in IT, there will be a long tail of organizations who continue to use firewalls or servers that are not compatible for the foreseeable future.

Securing networks at the near edge

Securing edge computing resources is crucial to protect sensitive data and ensure the integrity and availability of services. Unlike resources in the cloud, you cannot assume mitigations such as the AWS Nitro platform are in place to prevent, for example, a poison ARP/MAC spoofing attack across an Ethernet segment.

Identity and Access Management (IAM)

Implement robust IAM policies and practices to control access to edge computing resources. Ensure that only authorized individuals or systems have appropriate privileges to interact with the resources. Use strong authentication mechanisms such as **Multi-Factor Authentication** (**MFA**) and enforce the principle of least privilege to limit access rights to what is necessary.

Encryption

Implement end-to-end encryption for data transmission and storage in edge computing environments. Use industry-standard encryption algorithms and protocols to secure data in transit and at rest. Encryption helps protect data from unauthorized access or interception, especially in scenarios where edge devices communicate over public networks.

Secure communication protocols

Use secure communication protocols, such as **Transport Layer Security** (**TLS**), for secure communication between edge devices, gateways, and the cloud. TLS provides encryption and authentication mechanisms that protect data from eavesdropping and tampering. Ensure proper certificate management and use trusted certificates to establish secure connections.

Device hardening

Apply security best practices to harden edge devices and gateways. This includes regularly patching and updating firmware, disabling unnecessary services and ports, and configuring secure network settings. Employ **Intrusion Detection And Prevention Systems** (**IDS/IPS**) to monitor and mitigate potential security threats.

Network segmentation

Implement network segmentation to isolate and protect critical edge computing resources. Use **Virtual LANs (VLANs)** or **Software-Defined Networking (SDN)** techniques to create separate network segments for different types of devices and services. This helps contain potential security breaches and limit the lateral movement of threats within the network.

Monitoring and logging

Implement comprehensive monitoring and logging mechanisms to detect and respond to security incidents in real time. Monitor network traffic, device logs, and system events to identify any suspicious activities or anomalies. Employ centralized logging and analysis tools to gain visibility into edge computing resources and enable proactive threat detection and incident response.

Security updates and vulnerability management

Stay updated with security patches and firmware updates for edge devices and gateways. Regularly scan for vulnerabilities and implement a robust vulnerability management process to address any identified weaknesses promptly. Consider using automated tools for vulnerability scanning and patch management.

Physical security

Ensure the physical security of edge computing resources. Protect devices from unauthorized access, theft, or tampering by implementing appropriate physical security controls. This may include secure cabinets, access control systems, surveillance cameras, and other physical security measures.

Incident response and disaster recovery

Develop an incident response plan and disaster recovery strategy that's specific to edge computing environments. Define procedures for responding to security incidents, including containment, investigation, and recovery. Regularly test the effectiveness of the plan and conduct drills to ensure preparedness.

Training and awareness

Provide security training and awareness programs for personnel involved in managing edge computing resources. Educate employees about best security practices, social engineering threats, and the importance of following security policies. Promote a culture of security awareness throughout the organization.

By implementing these security measures, organizations can enhance the protection of their edge computing resources and mitigate potential risks. It is essential to continually evaluate and update security measures to stay ahead of evolving threats and ensure the ongoing security of edge computing environments.

Summary

In this chapter, we surveyed common pitfalls associated with the network and security facets of near-edge computing solutions. This included challenges faced on the internet due to latency, packet loss, and server and client configurations, as well as common protocols in use.

We also covered standard industry approaches to mitigating these issues. We explored how GSLB and IP Anycast are used to reduce the latency introduced by the physical distance between the server and the client. Then, we reviewed HTTP/3 and QUIC – a new set of protocols that eliminates the need to worry about many of the challenges faced by older, more widely adopted protocols such as HTTP/1.1 and HTTP/2.

Lastly, we covered some of the key considerations you must take into account regarding security when implementing an edge computing solution that is not fully based in the cloud.

In the next chapter, we will dive into the same sorts of details for solutions deployed at the far edge.

3

Understanding Network and Security for Far-Edge Computing

Edge computing in situations where reliable, high-speed internet access is not a given due to location or the nature of the devices involved are known as far-edge use cases. Examples include a mobile data center for disaster response, remote sensors for smart agriculture, a pilot station for a military UAV, or content delivery onboard a commercial airplane in flight.

In this chapter, we're going to cover the following main topics:

- Introduction to radio frequency (RF) communications
- Utilizing cellular networks
- Optimizing Wi-Fi (802.11x)-based connectivity
- Connecting to low-powered devices with LoRaWAN
- Integrating SATCOM in remote scenarios

Introduction to radio frequency (RF) communications

Almost all far-edge networking involves wireless networking of some type, and forms of wireless networking communicate using RFs. Wi-Fi, Cellular, LoRaWAN, SATCOM – they all use radio waves in a similar way as the radio in your car does.

Frequency and wavelength

Electromagnetic radiation consists of an electrical field that varies in magnitude, and a magnetic field oriented at a right angle to its electrical field. Both of these fields travel at the speed of light. Everything from cell phones, radio stations, and Wi-Fi access points to GPS signals are electromagnetic waves.

They are called electromagnetic because they are synchronized oscillations of electric and magnetic fields. But what does that mean? Take a look at the following figure. The part labeled E (the waves going up and down) represents the electric field and the part labeled B (the waves going left and right) represents the magnetic field:

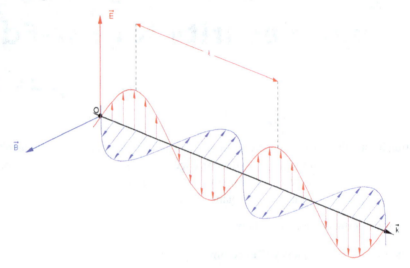

Figure 3.1 – An electromagnetic wave

The simplest way to generate an electromagnetic wave is to make a loop of wire and move a magnet back and forth within that loop. At its core, that's what's happening inside of a radio transmitter. It's just doing this at a microscopic scale, and oscillating very, very fast.

> **Frequency**
>
> Frequency is the number of wave cycles that pass a given point within a unit of time. It is often represented by the Greek letter Nu (n). It is measured in terms of **Hertz (Hz)**, with one hertz representing one cycle per second.

Kilohertz – KHz, 1000Hz, or one thousand times per second

Megahertz – MHz, 1,000,000Hz or one million times per second

Gigahertz – GHz, 1,000,000,000Hz or one billion times per second

Because these waves *always*[1] travel at the speed of light (*c*), there is a relationship between the frequency and the wavelength of any radio signal. The higher the frequency, the shorter the wavelength.

> **Wavelength**
>
> Wavelength is the physical length of an entire wave cycle. It is usually represented by the Greek letter lambda (λ). It is measured in terms of length such as meters, centimeters, millimeters, nanometers, and so on.

You can figure out the wavelength if you know the frequency and vice versa because of the following relationships:

$$v = \frac{c}{\lambda} \qquad\qquad\qquad \lambda = \frac{v}{c}$$

*λ represents the wavelength, n the frequency, and **c** the speed of light*

There are many calculators on the internet to help you make the conversion, but it is important to understand that these two things are tightly related.

Because *c* is so fast, 1Hz is a really low frequency and conversely would be a very long wavelength – about 186,282 miles (300,000 km). The only communication systems that get anywhere close to that (~50Hz) have to use the Earth itself as an antenna to generate long enough radio waves. They are called **Extremely Low Frequency** (ELF) and are used to communicate with submarines deep underwater – at data rates measured in a few bits per second. On the other hand, a 24GHz 5G signal that your mobile device receives only has a wavelength of 12.5mm – with achievable data rates of multiple gigabits/sec.

To recap, remember these general rules about frequencies:

- **Lower frequencies**: They have longer wavelengths, are effective over longer distances[2], and penetrate obstructions better[3], but have lower data rates

- **Higher frequencies**: They have shorter wavelengths, travel shorter distances[2], and are more affected by obstructions[3], but have higher data rates

---

1   The speed of light through a given medium varies according to that medium's refractive index. The amount of impact air has is small enough to ignore, while water slows electromagnetic waves down by as much as 25%. Light passing through a diamond slows by as much as 60%.

2   A more accurate way to explain this is to say shorter wavelengths travel less far before attenuation, or loss of field strength, makes them unusable.

3   Different kinds of materials affect electromagnetic signals differently. Not all obstructions are the same.

EM spectrum

Different communication technologies use different parts (also called *bands*) within the electromagnetic spectrum. Take a look at the following figure. You will notice that there is no fundamental difference between AM radio, Wi-Fi, visible light, X-rays, and even deadly gamma rays. The difference is simply one of frequency:

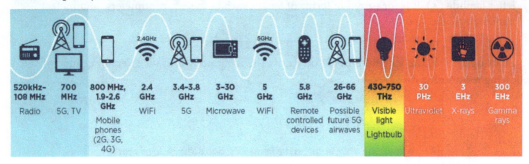

Figure 3.2 – Overview of how the electromagnetic spectrum is used

There's no need to worry, though. Your Wi-Fi signal operates at a frequency of 5 GHz, while visible blue light has a frequency of 750THz – over 150,000 times higher. Gamma rays are 200 million times higher. X-rays have wavelengths the size of an atom and gamma rays are the size of an atomic nucleus. It's a completely different ballpark. RF technologies are often referred to with an abbreviation, such as VHF or UHF. Those acronyms refer to the ITU frequency band specification they operate within.

Antennas

To know more about antennas we will cover the size, polarization, and types of antennas. We'll look at these three in detail in the following sub sections.

Size

An antenna's size is directly related to the wavelength of the signal involved – which, as you'll recall, is inversely proportional to the frequency. Higher frequency signals generate physically shorter waves. For optimal reception, the length of the antenna for both the transmitter and the receiver should match the physical size of the wave exactly.

For instance, FM radio broadcasts have a wavelength of around 9.8 feet (3 meters):

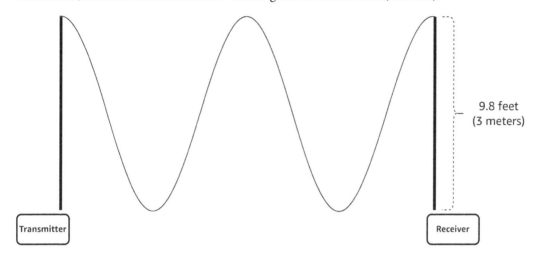

9.8 feet
(3 meters)

Transmitter

Receiver

Figure 3.3 – Optimal antenna size for a 100MHz FM broadcast

This isn't an issue for the transmitter, but few people are willing to drive around with such an antenna on their car. With AM radio wavelengths being measured in hundreds of feet, the problem is even worse.

Fortunately, one or both sides can achieve near-optimal reception with an antenna size that is ¼ the size (or another multiple) of the wavelength. Unidirectional broadcasts tend to have a full-sized antenna to optimize the transmission (wattage is money after all), but much smaller antennas on the receive end.

Keep in mind that we're talking about optimal reception. If the receive antenna on your car is off by a few inches, the relatively small amount of signal loss is made up for with signal processing techniques inside the receiver.

Polarization

Polarization can be divided into two main categories. Let's look at them in detail.

Linear

So far, we've been talking about a simple case where the antenna matches on both sides in terms of size. Now, let's look at what happens when they are not the same in terms of orientation:

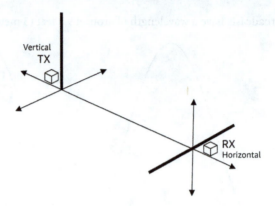

Figure 3.4 – Misaligned monopole antennas

In this case, the receiving station will not be able to "catch" the radio waves out of the air because it is completely perpendicular to the transmitting station. While it's okay to be a little bit misaligned, the farther out of alignment one side gets, the worse the reception.

If you are old enough to remember moving the rabbit ears around on a TV set until a particular channel came in more clearly, this helped because you were re-orienting the antenna on your receiver to more closely match that of the antenna on the roof of the relevant TV station.

This orientation of the signal is known as its polarization. So far, we've only been talking about what is known as linear polarization – the signal is straight up and down (vertical) or left to right (horizontal).

Circular

Sometimes, the antennas do not remain in a fixed orientation and rotate continually – sometimes at random. This is also known as the Faraday rotation. It is a common problem when communicating with satellites in space. Circular polarization is a method that can be used to overcome it:

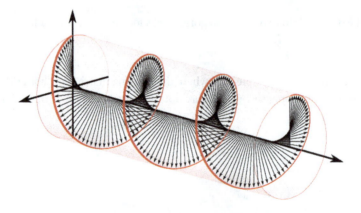

Figure 3.5 – Circular polarization of a radio signal

This can be thought of as putting a spin on the signal as it leaves the transmitter. This makes the precise orientation of the receiving antenna less important but does increase the complexity of the antenna's design.

Types of antennas

Several types of antennas can effectively handle circular polarization. Let's take a look.

Helical

Helical antennas mimic the shape of the signal directly. Larger ones are obvious, but many are hidden beneath rubber or plastic covers:

Figure 3.6 - Examples of helical antennas designed for circular polarization

Microstrip

Also known as patch antennas [4], these consist of flat metal pieces of various sizes and shapes on a **Printed Circuit Board** (PCB). These can use either linear or circular polarization, depending on their design [5]:

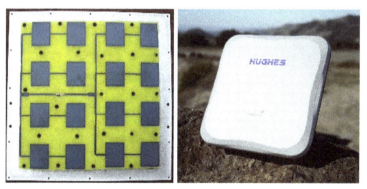

Figure 3.7 – A 2.4GHz patch antenna and an L-band SATCOM terminal

---

4   This is technically inaccurate, as a microstrip antenna consists of multiple patch antennas on the same PCB that work together as an array

5   Subtypes include the planar inverted-F antenna (PIFA) used in mobile devices

Advantages of this type include their low profile and small size. They are used in mobile phones, portable SATCOM terminals, Wi-Fi adapters in laptops, and similar applications where an external antenna is undesirable.

Parabolic

Most people associate parabolic antennas with satellite communications. However, parabolic dishes aren't themselves the antenna. They are a method to increase the gain of a feed antenna by reflecting signals and focusing them onto it. This feed antenna can be polarized in any way, hence why there are parabolic antennas that have circular polarization and others that have linear polarization.

Modulation

To transmit information using a radio wave, we need to modify that wave somehow to encode our data. There are three aspects of a signal that can be modulated:

1. **Amplitude-shift keying (ASK):**

    Modulation is based on the power or intensity of the signal. AM radio uses this method:

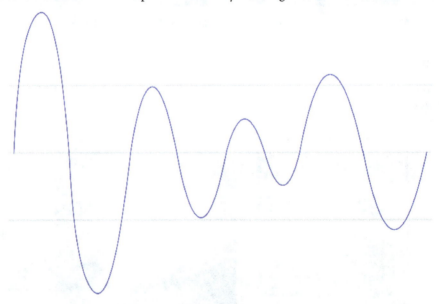

Figure 3.8 – Amplitude modulation

2. **Frequency-shift keying (FSK):**

Modulation is based on how often the wave repeats. FM radio uses this method:

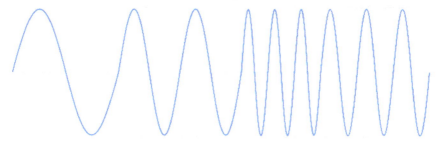

Figure 3.9 – Frequency modulation

3. **Phase-shift keying (PSK):**

Modulation is based on where in the cycle the wave is with respect to time. Wi-Fi and cellular technologies use this method:

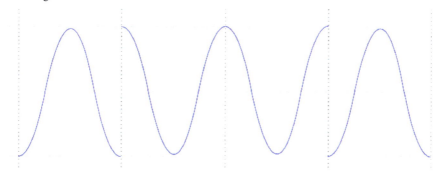

Figure 3.10 – Phase modulation

Duplexing

So far, we've discussed receive-only applications such as the AM/FM radio in your car, or the satellite TV dish on the side of your house. Wireless technologies that involve bidirectional communication must implement some form of duplexing. This can be thought of as the rules for whose turn it is to speak and whose turn it is to listen – or in some cases, how can both members of the conversation speak and hear simultaneously?

All duplexing techniques fall under one of two categories:

- **Half-duplex**: Only one party to the conversation may speak and the other must listen. Then, the roles are reversed. This can be based on the time or urgency of the messages involved.

- **Full-duplex**: Both parties can speak and hear at the same time. This can be accomplished by having separate "paths" set aside for each function, or more complex approaches.

Frequency division duplexing (FDD)

FDD is a full-duplex technique whereby the bandwidth of a given channel is split between transmit and receive sub-channels. Historically, this has been the most common method of achieving bidirectional communication over wireless technologies. It is the preferred method in lower frequency bands:

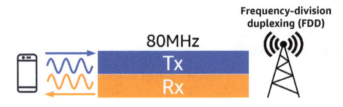

Figure 3.11 – FDD

FDD is the simplest approach. Because of that, it is also the cheapest to implement in hardware. Unfortunately, in most implementations, it is not possible to dynamically increase or decrease the transmit and receive channels to reflect the needs of a given endpoint. Think of a mobile device that keeps the RX channel utilized at 90% when downloading web pages, but the TX channel is barely touched at 5% because the requests for those pages are small. Or that person you know who monopolizes the conversation and never lets you talk.

Time-division duplexing (TDD)

TDD is a half-duplex technique that simulates full-duplex by switching between uplink and downlink phases on the same frequency at regular intervals:

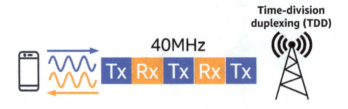

Figure 3.12 – TDD

The main advantage TDD has is that it halves the frequency range required versus FDD. However, it brings a couple of disadvantages with it. It requires precise time synchronization on both ends – on the order of milliseconds. This increases cost and complexity.

For voice calls or low-throughput data streams such as SMS, simple TDD was fine for a long time. Eventually, however, users started doing things such as video streaming to their mobile devices. People in the industry started eyeballing the bandwidth that was being wasted by a straight 50/50 approach.

Dynamic time-division duplexing (D-TDD)

D-TDD is a half-duplex approach that utilizes the same principles as static TDD while adding a mechanism that adjusts how often downlink and uplink phases occur. This is achieved through the use of a special control message often called a **Slot-Format Indicator (SFI)**. This sets the schedule, so to speak, of the next 14 phases. The arrangement of upload and download phases can be skewed in either direction dynamically based on the needs of the system at that moment in time.

In the following figure, each slot is broken into 14 symbols, and each symbol can be of the uplink, downlink, or flexible type. The arrangement of these symbols is governed by the SFI on the left. The SFI that the stations should use can be exchanged using any number of control-channel mechanisms. The specifics of this exchange depend on the technology in question:

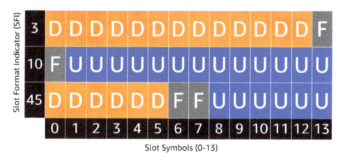

Figure 3.13 – D-TDD

It isn't perfect, but on average, it reduces the wastefulness of static TDD by about 75%. The most obvious disadvantage is complexity – which leads to increased costs for both the hardware and ongoing management.

Multipath propagation

This is a classic problem that's faced by electromagnetic-based technologies of all types. It refers to when the same signal sent by the transmitter is received on the other end via two or more paths. This can be caused by the electromagnetic waves being reflected off man-made objects such as buildings and airplanes, as well as by mountains or activity within the atmosphere itself:

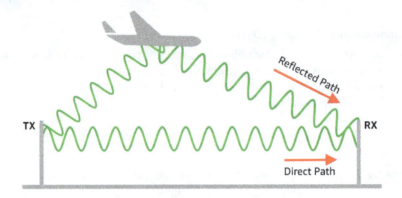

Figure 3.14 – Secondary transmission path created by reflection

This is a problem because those two paths are of different lengths. That means the radio waves will arrive offset from each other. This is also known as being *out of phase*:

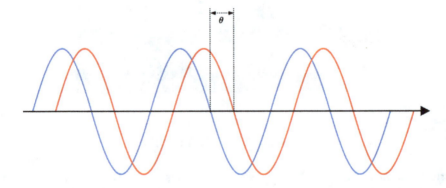

Figure 3.15 – The same signal from different paths will be phase-shifted

This is particularly problematic for technologies such as GNSS, where precision timing of signal arrival is a fundamental assumption.

Multiple input multiple output (MIMO)

MIMO is a method of increasing the effective capacity of a radio link by deliberately exploiting multipath propagation. This is typically accomplished via the use of multiple transmitters and receivers on both sides [6]:

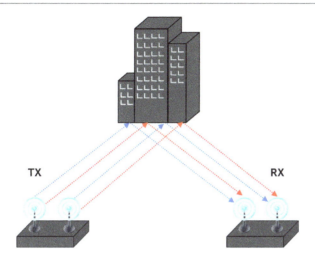

Figure 3.16 – MIMO exploiting multipathing via reflection

Multiplexing

Duplexing is about splitting up a single conversation between transmit and receive functions – but the two people talking on the channel are the same. Multiplexing, on the other hand, is about bundling up many two-party conversations so that they can all share one link. Many networks have an expensive point of aggregation that many users need to share for efficiency. Undersea cables between countries or point-to-point microwave links across a city are some examples.

Occasionally, multiplexing is something that gets added to a technology long after the original transmission medium was thought to be fully utilized. **Digital Subscriber Line (DSL)** is an example of this. Telco customers started demanding broadband internet access long before fiber optic cables were run to their neighborhoods, let alone to their houses. In many areas, those old copper telephone lines for the **Plain Old Telephone service (POTS)** network were run to the same places decades ago. DSL is a technique that multiplexes an additional analog signal onto the old POTS phone lines, but because it is outside the range of human hearing, you could still make a phone call and never notice.

---

6   Some implementations have a single antenna on one side and multiple antennas on the other.

Wave division multiplexing (WDM)

WDM is multiplexing with channelization based on wavelength. It often corresponds to colors of visible light when used on fiber-optic networks.

Consider a phone call from New York to London. At some stage, your conversation is getting funneled to an undersea fiber-optic cable traversing the Atlantic Ocean. Lots of individual phone calls happen at once over that big fat pipe, yet they do not interfere with each other. The voice call is converted into a specific color of laser light by a transponder and sent, along with many other colors, across the same fiber-optic cable by a device called a *muxer*. On the other end, a *demuxer* distributes the colors to transponders, which convert the signal back to voice:

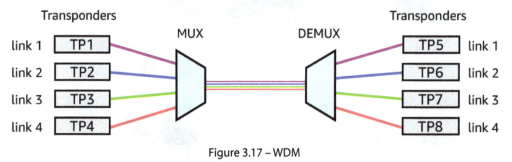

Figure 3.17 – WDM

Time-division multiplexing (TDM)

TDM is a method of transmitting multiple digital signals over a single channel or link by dividing the time available for transmission into distinct time slots. Each signal is assigned a specific time slot, and the signals are transmitted sequentially, one after the other. This allows multiple signals to share the same physical channel, effectively increasing the capacity of the channel.

TDM is often used in telecommunications to transmit multiple phone calls over a single telephone line, allowing multiple conversations to take place simultaneously. It is also used in other types of communication systems, such as broadband networks and data centers, to transmit multiple data streams over a shared connection.

In TDM systems, the time slots are typically assigned using a specific pattern, such as a fixed schedule or a round-robin approach, to ensure that each signal gets an equal share of the available time. TDM is generally less efficient than other multiplexing techniques, such as WDM, but it is easier to implement and can be used in a wide range of applications.

Orthogonal frequency division multiplexing (OFDM)

This is a scheme in which multiple closely spaced *orthogonal* subcarrier signals with overlapping spectra are transmitted to carry data in parallel.

In traditional **Frequency Division Multiplexing** (**FDM**), the subcarriers (also known as *channels*) are kept apart using a little bit of space between them called a guard band:

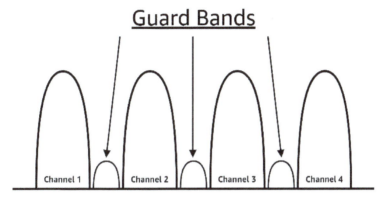

Figure 3.18 – FDM using guard bands

This is done to prevent crosstalk, noise, or interference between the channels. It also makes it easier for the demodulators to single out the channels when demuxing them.

OFDM deliberately overlaps the channels in a specific way – this is where the orthogonal part comes in. Orthogonal means "at right angles," but in this context, it refers to a precise mathematical relationship between how the channels are spaced across the frequency band. This technique can save as much as 50% of the bandwidth, which can now be used to carry additional channels:

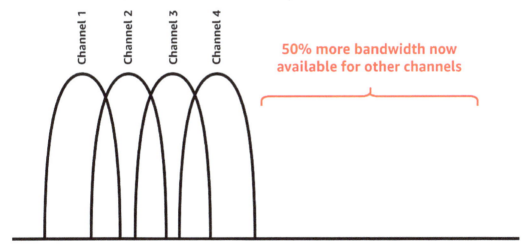

Figure 3.19 – OFDM

OFDM uses digital signal processing techniques to perform *coherent demodulation* on these overlapping channels [7]. The mathematics are beyond the scope of this book. At its core, this is simply another example of how we can exploit the fact that light always travels at the same speed. OFDM techniques can, and often are, used in combination with MIMO.

Shannon-Hartley theorem (signal-to-noise ratio)

Developed in the 1940s, the Shannon-Hartley theorem describes the maximum rate at which information can be transmitted over a communications channel of a specified bandwidth in the presence of noise:

$C = B \log 2(1 + S\_N)$

Here, we have the following:

C is the channel capacity in bits per second

B is the bandwidth of the channel in hertz

S is the average received signal power in watts

N is the average power of noise/interference in watts

Let's zero in on the two most important terms to remember:

- **Signal (S)**: Average power of the received signal in watts
- **Noise (N)**: Average power of noise (that is, interference) in watts

These two terms are grouped into a single expression known as the **Signal-to-Noise Ratio** (**SNR**). An SNR of 2:1 means there is twice as much signal as there is noise. An SNR of 1:1 means there is the same amount of noise as there is a signal.

Another way to put this is to say a signal that is suffering throughput loss from degradation due to interference can be improved by increasing the signal's power [8].

Utilizing cellular networks

In this main section we will take a closer look at the different cellular networks and understand how we can utilize them. We will mainly cover 4G/LTE, 5G, C-V2X, and NB-IoT.

---

7   Fourier transforms can be performed to convert the time domain of a digital signal's square waves into frequency domains corresponding to the channels.

8   Keep in mind that increasing the power of your signal can create interference for others. This is why there are often laws limiting how powerful a given device's transmitter is allowed to be.

4G/LTE

What is known as 4G/LTE is not a single specification. It is a family of technologies that set out to meet a proposed definition of 4G laid out by the ITU in 2008. Its designers had the following improvements in mind over 3G:

- Fully packet-switched (3G was circuit-switched)
- Peak data rates up to 100 Mbps for mobile devices
- 1 Gbps for stationary devices such as 4G hotspots

Increased density of devices per cell through resource sharing:

Figure 3.20 – An example of a 4G/LTE network

How 4G/LTE is implemented varies considerably between **Mobile Network Operators** (**MNOs**). There are also key differences in how a given MNO's 4G/LTE network functions across regions[9]. MNOs began rolling out 4G/LTE networks around 2011, and it was 2016 before MNO coverage could be considered widespread.

Evolved Node B (eNodeB)

The part of a 4G/LTE network you are likely most familiar with is the front end – the ubiquitous cell tower. In 4G/LTE parlance, these are known as eNodeBs. They are elements of a standard cellular network component known as the **Radio Access Network** (**RAN**). They include antennas, transceivers, and radio access controllers.

Evolved Packet Core (EPC)

Note that 4G/LTE base stations (eNodeBs) only communicate with each other directly for control plane functions, such as to hand off a device from one tower to another.

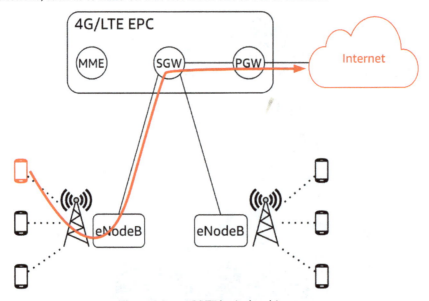

Figure 3.21 – 4G/LTE logical architecture

Otherwise, communication needs to go through one of the subcomponents of EPC:

- **Serving gateway** (**SGW**): Routes user data plane traffic, either between mobile devices or out to other EPC functions, such as a **Packet Data Network Gateway** (**PGW**). It also provides core network services such as routing, switching, and transport of data packets.

---

9   The Americas, Europe, Africa, and Asia all had different regulatory constraints that drove this.

- **Packet data network gateway** (**PGW**): Routes user data-plane traffic between EPC and external IP networks such as the internet. It's also responsible for handling the exchange of data between the mobile device and the wider internet, and it consists of several interconnected network elements.

- **Mobility management entity** (**MME**): This handles critical control plane functions for mobile devices, including authentication, location tracking, and handover signaling.

4G/LTE latency

EPC instances are centralized and often physically distant from the eNodeBs in a cellular network.

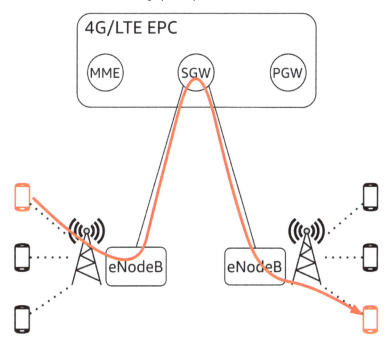

Figure 3.22 – 4G/LTE hairpin routing

Because user data plane traffic has to go back up to the EPC layer to be routed (this is known as *hairpin routing*), the average RTT on 4G hovers around 50 ms.

5G

As of August 2021, 175 MNOs were operating public 5G services across 72 countries[9]. It is estimated that 5G networks will account for 77% of MNO revenues (600 billion USD) by 2026, with demand for both consumer and business services such as MEC driving adoption. Much of this is being driven by the massive deployment of cellular-connected IoT devices, which are predicted to top six billion by

2026. That will be the point where IoT devices overtake smartphones as endpoints on mobile networks, with half of these expected to use 5G connections.

5G benefits from widespread support as a single global standard. When the specification was developed, the primary design goals were as follows:

- Peak data rates up to 10 Gbps

- Reliable, deterministic low latency for critical applications

- Much higher density of devices on the network

- **Network Functions Virtualization** (**NFV**) capabilities built into the core

- Ability to fine-tune **Quality of Service** (**QoS**) parameters per application

At the same time, they realized that MNOs had made considerable investments in 4G/LTE infrastructure. Therefore, the specification was formulated in such a way that brand-new end-to-end 5G networks were not a requirement. Deployments are typically done in a phased manner that allows elements of an MNO's network to be upgraded over time:

Figure 3.23 – Example of a 5G network

Even where standalone/private 5G networks are built using the full **5G New Radio** (**5G NR**) architecture end-to-end, **User Equipment** (**UE**) such as mobile devices themselves are often built in a hybrid way such that 4G/LTE acts as a fallback position in case of incompatibilities.

5G Core (5GC) architecture

5G Core (**5GC**) is the basis of the network architecture used in 5G (fifth-generation) mobile networks. It is responsible for providing the same core network services as EPC, but it has been redesigned to support the increased demands and requirements of 5G networks. Compared to 4G/LTE EPC, 5GC was designed to be more flexible and scalable, with the ability to support a wider range of use cases and network architectures.

5GC includes the following key elements:

- **Access and mobility management function** (**AMF**): Manages authentication, radio resource management, handover management, connectivity to external networks, and management of QoS for user data plane traffic.

- **Session management function** (**SMF**): Manages the establishment, maintenance, and termination of sessions between the mobile device and the network.

- **User plane function** (**UPF**): Routes user data plane traffic between the mobile device and the network. It is also responsible for compressing packets and enforcing the QoS policies set by the AMF:

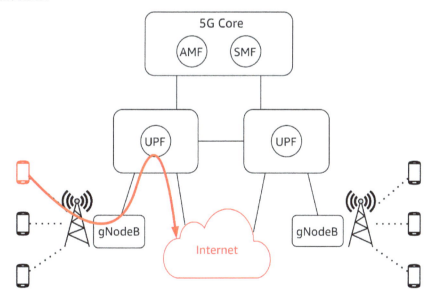

Figure 3.24 – 5GC logical architecture

In the preceding figure, we can see that, unlike the PGW in 4G/LTE EPC, there is no longer a single node acting as the gateway to the internet or other packet networks. Mobile devices no longer need to backhaul to the PGW to leave the cell provider's network. Elimination of this bottleneck was needed to support the much higher density of mobile devices, which is a key use case for 5G:

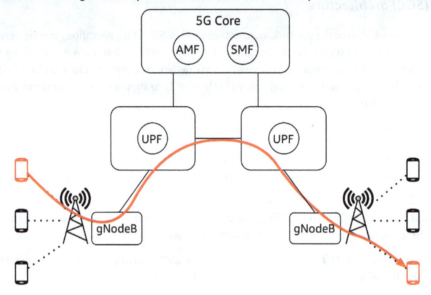

Figure 3.25 – 5G intranetwork routing

The preceding figure illustrates how the data path between two mobile devices on the same network benefits from 5GC's distributed UPF. These changes are a key reason 5G devices see average RTTs of <10ms versus the average of 50ms observed in 4G/LTE networks.

Network slicing

Network slicing is a technique in 5G that can be thought of as a combination of VLANs and QoS mechanisms seen in enterprise data networks. Some aspects of them could be looked at as analogous to VPCs and SGs in AWS.

Regardless of how you conceptualize them, 5G slices allow multiple virtual networks to coexist on the same physical infrastructure. This allows for very fine-grained control of security and performance parameters down to a per-slice basis. MNOs often have the average user on general use public slices, while carving off per-customer slices for their B2B customers. Sometimes, mobile devices are given access to multiple slices from one device, each one mapping to a different application.

The **Third Generation Partnership Project (3GPP)** has defined three network slice categories:

- **Enhanced Mobile Broadband (eMBB)**: Designed to ensure high data rates to mobile devices, with SLA targets of >100 Mbit/s average and >10 Gbit/s peak throughput.

- **Ultra-Reliable Machine Type Communication** (**uMTC**): Focuses on the reliability and deterministic latency aspects of 5G. SLAs target 3 9's service availability and <1ms RAN latency. Sometimes, this is called **Ultra-Reliable Low-Latency Communication** (**URLLC**).

- **Massive Machine Type Communication** (**mMTC**): Concentrates on the density of devices with lots of small conversations. This is also known as **massive Internet of Things** (**mIoT**).

3GPP also defines dozens of application-specific network slice templates such as those for all subcategories of V2X. In addition to these standard categories, MNOS can engineer custom slice types in response to customer demand.

Network function virtualization (NFV)

NFV uses proven hypervisor and/or container platforms to eliminate the 1:1 mapping between hardware and function that was seen in 4G/LTE EPC. 5G components, on the other hand, are deployed as virtual machines or containers on commodity compute hardware:

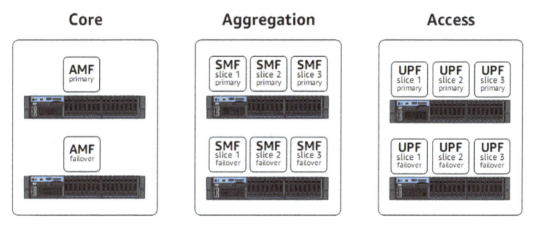

Figure 3.26 – 5G functions via NFV on commodity servers

This allows 5G service providers to deploy, manage, and scale the critical components of their network in an automated way. This not only reduces cost and time-to-market, it improves reliability and SLA adherence – which are critical to an MNO's business.

While NFV was possible in 4G/LTE EPC, 5GC was built from the ground up with it in mind. All functions of 5GC can be virtualized – AMF, SMF, UPF, and network slicing can all be deployed as virtual constructs from the 5G management plane and operated transparently by the 5G control plane.

Small cells

So far, we have been discussing macrocells. They are large arrays of antennas that are typically mounted on their own tower and meant to service all of a CSP's customers for a radius measured in kilometers.

The ever-growing demand for new mobile devices has driven a market in *small cells*. These are small, lower-powered access nodes that are deployed for specific uses.

CSPs add small cells to their existing networks to increase coverage in rural areas, to service more devices in an area of particularly dense usage, or to provide service indoors. Small cells are also found in most private 5G networks. Small cells are broken up into femtocells, picocells, and microcells – each of which has a different range and supports a different number of users.

5G frequency spectra

Unlike 4G/LTE, 5G frequencies are split into three range groupings, each in a different region of the spectrum:

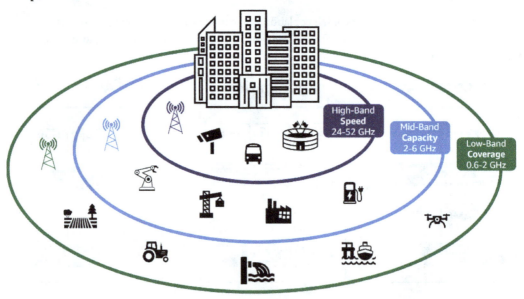

Figure 3.27 – 5G frequency band utilization

Cellular Vehicle-to-Everything (C-V2X)

Vehicle-to-Everything (**V2X**) is a set of specifications that encompass multiple types of wireless communication between a vehicle and its surroundings. This includes other vehicles, infrastructure, networks, and even pedestrians. V2X communication has the potential to revolutionize transportation, making it safer, more efficient, and more sustainable. C-V2X, however, is based on 5G (although it can use 4G/LTE in a more limited fashion).

V2X can be broadly categorized into four subtypes:

- **Vehicle-to-Vehicle (V2V)**: This communication occurs between vehicles on the road, allowing them to exchange information about their position, speed, and direction. This enables **advanced Driver-Assistance Systems (ADASs)** to prevent collisions, optimize navigation, and facilitate cooperative driving.

- **Vehicle-to-Infrastructure (V2I)**: In this type of communication, vehicles interact with roadside infrastructure such as traffic signals, road signs, and smart city sensors. This allows for real-time traffic management, improved safety measures, and enhanced navigation guidance. China is leading the way in this area. Nearly 90 cities have already partnered with local wireless network operators, deploying tens of thousands of roadside units to demonstrate intelligent highways and urban intelligent networked roads.

- **Vehicle-to-Pedestrian (V2P)**: This type of communication occurs between vehicles and pedestrians or cyclists, using devices such as smartphones or wearable technology. V2P communication can help prevent accidents by providing alerts to both pedestrians and vehicle drivers about potential collisions.

- **Vehicle-to-Network (V2N)**: This type of communication connects vehicles to various networks, including the internet, cellular networks, and cloud-based services. V2N communication can provide vehicles with updates on traffic, weather conditions, and other relevant information to enhance their performance and safety.

Of note is that, unlike most other 5G technologies, C-V2X does not necessarily require an MNO's infrastructure to function. It can operate without a SIM, without network assistance, and uses GNSS as its primary time synchronization source. Today, about 50-60% of vehicles in North America are equipped with a cellular modem. The decision-making process within the automotive industry on whether to standardize on DSRC/802.11p or 5G for V2X has been long and drawn out but has finally settled on using cellular as the standard going forward.

According to the **5G Automotive Association (5GAA)**, auto manufacturers that are currently producing C-V2X capable models include Audi, BMW, Daimler, Ford, Lexus, Nissan, and Tesla.

Narrow-Band IoT (NB-IoT)

NB-IoT is a specification devised by 3GPP that defines a **low-powered WAN (LPWAN)** technology that rides on top of existing 4G/LTE and 5GC networks. It is meant to provide a lower cost level of service for IoT devices that do not need the full throughput of an MNO's standard 4G/LTE or 5G data service offering.

Because it piggybacks on top of existing mobile networks, it shares the same licensed frequency spectrum, and normally the same cell towers/antennas. However, at a signal level, it functions a bit differently. The specification limits each device to a maximum of 200KHz of bandwidth. Contrast this with 4G/LTE, which can have 20MHz channels, and 5G, which can go as high as 400MHz, and the reason it is called "narrow-band" becomes evident. An MNO can support as many as 100 NB-IoT devices using the same amount of bandwidth needed to support a single 4G/LTE phone using a 20MHz channel.

How much throughput an NB-IoT device can squeeze out of that 200KHz channel depends on the version. 3GPP Release 17 was published in 2022 and specifies the latest revision, known as NB-IoT Enhanced. This version specifies a maximum throughput of 250 kbps down and 20 kbps up. It achieves this by using TDD to time-slice the transmit phase as FDMA and the receive phase as OFDMA.

Another difference is that NB-IoT is typically deployed using the guard band slots of an MNO's network. While this is not always true, it is important to ask your MNO whether they deploy NB-IoT using "in-band mode" or "guard-band mode" as the latter will inevitably suffer from a higher signal-to-noise ratio than you could expect from an NB-IoT channel provisioned in a standard slot. Guard bands exist for a reason. At the time of writing, few NB-IoT offerings do not use guard-band mode:

Figure 3.28 – NB-IoT-capable pressure sensor

In most other ways, NB-IoT works like any 4G/LTE or 5G mobile device. Each device needs a SIM (although eSIMs are becoming the standard) to access the MNO's network. Each device is also paired with one cell tower/radio at a time. Finally, the connection is synchronous, which means it is constantly on, regardless of whether the device has data to send or receive.

The narrowness of the band allows the MNO to charge less for the service, but it also means NB-IoT devices need less power for the transceiver than if they were using standard 4G/LTE or 5G. However, because of the synchronous connection, NB-IoT devices as a rule consume more power than LPWAN technologies that use an asynchronous connection model.

Optimizing Wi-Fi (802.11x)-based connectivity

Wi-Fi was designed to allow laptops, smartphones, and tablets to connect to the internet and/or communicate with each other on a **local area network** (**LAN**). It uses RF to transmit data over relatively short distances, typically within a home or office – although permutations intended for outdoor use are becoming more common.

Wi-Fi is based on the IEEE 802.11 standards, which operate at Layer 1 of the OSI model (physical). Introduced in the late 1990s, it was the first commercially successful wireless networking technology that was designed to work seamlessly with Ethernet (IEEE 802.3) – which almost all LANs use at Layer 2.

Wi-Fi-1 through Wi-Fi-6

The following table shows us the comparison of 802.11a/b/g/n/ac/ax:

	802.11 (b) Wi-Fi-1	802.11 (a) Wi-Fi-2	802.11 (g) Wi-Fi-3	802.11 (n) Wi-Fi-4	802.11 (ac) Wi-Fi-5	802.11 (ax) Wi-Fi-6
Max Speed	11 Mbps	54 Mbps	54 Mbps	600 Mbps[10]	1.3 Gbps[11]	1.7 Gbps[12]
Range Indoor (2.4)	35 m	N/A	45 m	60 m	N/A	60 m
Range Indoor (5)	N/A	30 m	30 m	45 m	45 m	45 m
Range Outdoor (2.4)	70 m	N/A	90 m	120 m	N/A	120 m
Range Outdoor (5)	N/A	60 m	75 m	90 m	90 m	90 m
2.4 GHz Band	Yes	No	Yes	Yes	No	Yes
5 GHz Band	No	Yes	Yes	Yes	Yes	Yes
OFDM	No	Yes	Yes	Yes	Yes	Yes
MU-OFDMA	No	No	No	No	No	Yes

---

10   Requires the use of vendor-specific proprietary beamforming/spatial streams.

11   Refers to per-station throughput. The whole network theoretical maximum is 6.9 Gbps.

12   Refers to per-station throughput. The whole network theoretical maximum is 9.6 Gbps.

	802.11 (b) Wi-Fi-1	802.11 (a) Wi-Fi-2	802.11 (g) Wi-Fi-3	802.11 (n) Wi-Fi-4	802.11 (ac) Wi-Fi-5	802.11 (ax) Wi-Fi-6
SU-MIMO	No	No	No	Yes	Yes	8x8
MU-MIMO (d)	No	No	No	No	4x4	8x8
MU-MIMO (u/d)	No	No	No	No	No	8x8
Spatial Streams	No	No	No	No	4	8

Figure 3.29 – Comparison of 802.11a/b/g/n/ac/ax

Modulation and coding schemes (MSCs)

The speeds provided in the preceding table are best-case scenarios. They assume an optimal SNR, which, in turn, allows the use of a modulation and encoding scheme that gets a higher data rate. Each generation of Wi-Fi has a different matrix of MCSs. The following is the MCS index table for 802.11ac (Wi-Fi-5):

	Modulation	FEC Coding Rate	Data Rate
MCS0	BPSK	1/2	
MCS1	QPSK	1/2	2x faster than MCS0
MCS2	QPSK	3/4	3x faster than MCS0
MCS3	16-QAM	1/2	4x faster than MCS0
MCS4	16-QAM	3/4	6x faster than MCS0
MCS5	64-QAM	2/3	8x faster than MCS0
MCS6	64-QAM	3/4	9x faster than MCS0
MCS7	64-QAM	5/6	10x faster than MCS0
MCS8	256-QAM	3/4	12x faster than MCS0
MCS9	256-QAM	5/6	13.3x faster than MCS0

Figure 3.30 – 802.11ac modulation and coding schemes

Each of the MCSs shown has two parameters:

- **Modulation**: In this context, modulation refers to the particular 802.11x modulation type in use. Some modulation types are very sensitive to noise while others tolerate it well. However, the robustness of a modulation type is achieved by reducing how sensitive it is – and this means a lower bit rate.

- **FEC coding rate**: This describes how many bits transfer data, and how many are used for forward error correction. A coding rate of 5/6 means for every 5 bits of useful information, the coder sends 6 bits of data. In other words, there's one error bit for every 5 data bits:

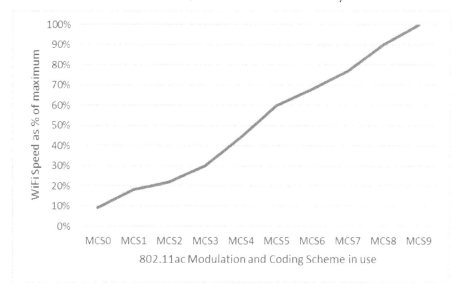

Figure 3.31 – Impact of MCS on data rate for 802.11ac

A Wi-Fi-5 or Wi-Fi-6 access point will negotiate the best MCS that it can, given the interference it is experiencing. Wi-Fi devices tend to express the SNR as a single number in dB, which represents the amount of signal above whatever noise is present.

A laptop 1 meter away from an access point with no obstructions would have an SNR of ~50 dB, and be able to operate at MCS9 (100% max speed). A second laptop far away or in a different room might only see an SNR of ~25 dB and be stuck at MCS3 (30% max speed).

Here are some practical steps that can help your device negotiate a faster MCS to its access point:

- **Reduce devices per AP**: Try to have only 3-4 devices per AP where possible

- **Change Wi-Fi channels**: Utilities such as NetSpot can help with this

- **Increase AP signal power**: Some APs default to a lower power level than they are legally able to use

Spatial streams

The term for beamforming as it is implemented within Wi-Fi is *spatial streams*.

While some vendors of 802.11n (Wi-Fi-4) devices did implement beamforming, it was through proprietary mechanisms that were specific to each product line. 802.11ac (Wi-Fi-5) was the first to include it as part of the specification.

When a Wi-Fi access point has beamforming enabled, it first estimates the *angle of arrival* of each client by comparing small differences in arrival times of a signal across multiple antennas that are close together. Once it knows the direction in which it needs to steer the beam, it will have those antennas broadcast the signal at slightly different times. The pattern that's used is known as a *steering matrix*.

This deliberately introduces interference because the waves now overlap a little bit. However, not all interference is the same. Some are *constructive interference*, which makes the signal stronger in one direction, while *destructive interference* makes it weaker in another:

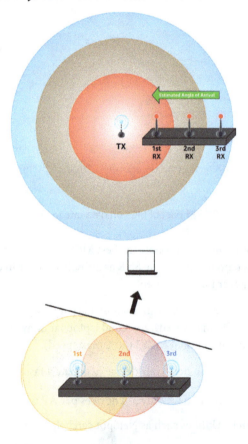

Figure 3.32 – Beamforming with 802.11ac (Wi-Fi-5)

The net effect of all this is to maximize the signal strength on a *per-client* basis. This means the signal effectively travels farther and penetrates obstacles better. With older Wi-Fi specifications, all you could do is increase the power output of an omnidirectional signal or add Wi-Fi repeaters.

This is one of the reasons for a seemingly endless multiplication of antennas on even consumer-grade access points. More antennas on both the AP and the clients are better for Wi-Fi throughput – up to a point[13]. Regardless of the number of antennas, the 802.11ac (Wi-Fi-5) specification supports a maximum of four spatial streams to be active at once.

802.11ax (Wi-Fi-6) increased this to eight and also enhanced it by including client-side modifications that help the AP figure out where a given client is instead of leaving all the work on the AP.

WiFi and MIMO

As discussed previously, MIMO is a method for increasing effective throughput by deliberately exploiting multipath propagation. The different generations of WiFi make use of this in varying ways.

802.11n (Wi-Fi-4)

This supported the more limited **Single User MIMO (SU-MIMO)**. As its name suggests, SU-MIMO means the access point can only be sent to one client at a time.

802.11ac (Wi-Fi-5)

This added MU-MIMO (d). The (d) stands for downlink. With MU-MIMO (d), only one station can transmit, but multiple stations can receive at any given time.

802.11ax (Wi-Fi-6)

This was extended to MU-MIMO (u/d). Now, multiple devices can both transmit and receive simultaneously.

MU-OFDMA

Basic OFDM has been supported since 802.11a (Wi-Fi-2). 802.11ax (Wi-Fi-6) has extended this to now support multiple users.

You could think of the older style of OFDM as a sequence of trucks, each delivering boxes from one vendor at a set time every day. MU-OFDMA allows each truck to be loaded with multiple vendor's boxes. It also allows the delivery schedule of those trucks to happen only when there's a full load.

Older Wi-Fi specifications were designed for web browsing and checking email. Congestion emerged as video streaming, AR/VR, and gaming became common. This, combined with more and more client devices transmitting at the same time, meant that the queuing caused by simple OFDM increased latency.

---

13   Two antennas are the minimum for beamforming to function at all, while three is recommended.

Perhaps most importantly, MU-OFDMA allows priorities to be set not only per client but per protocol/traffic type. In other words, the access point could prioritize video streaming at one level, IoT messages at another, and mission-critical VOIP at the highest.

802.11p (DSRC)

An amendment to the broader IEEE 802.11 **Wireless LAN (WLAN)** standard, 802.11p is tailored for high-speed, short-range communication in a vehicular environment. The standard operates in the 5.9 GHz frequency band and utilizes the **Dedicated Short-Range Communications (DSRC)** protocol to ensure low latency and reliable data exchange.

The primary advantage of DSRC over 4G/LTE or 5G for V2X is that it can provide some value in the absence of any infrastructure. If two V2X-equipped cars come within range of each other, they will exchange information in a peer-to-peer fashion. This would function even in the middle of the Sahara.

In 2016, Toyota became the first automaker to introduce cars equipped with V2X systems, followed by GM in 2017. Both of these used DSRC as opposed to 4G/LTE or 5G. While DSRC was the first standard the automotive industry adopted, that is changing for several reasons. Compared to 4G/LTE or 5G for V2X, DSRC suffers from the following limitations:

- **Limited capacity and scalability**: DSRC operates in a narrow frequency band (5.9GHz), which limits its capacity to support a high number of simultaneous connections in dense traffic scenarios. 5G offers broader bandwidth and improved spectral efficiency, allowing it to handle more devices and users concurrently.

- **Lower data rates**: DSRC offers lower data rates compared to 5G, which hinders its ability to support advanced V2X applications that require higher throughput, such as high-definition video streaming for autonomous vehicles. 5G, with its enhanced data rates, can better accommodate these demanding use cases.

- **Latency**: Although DSRC provides relatively low latency communication, 5G has the potential to achieve even lower latencies, especially with the implementation of 5G **Ultra-Reliable Low-Latency Communication (URLLC)**. URLLC can enable mission-critical applications and real-time control systems that demand near-instantaneous response times.

- **Network slicing**: 5G supports network slicing, a feature that allows the creation of virtual networks tailored to specific use cases or applications. This enables the allocation of dedicated resources for V2X communications, ensuring the desired performance levels. DSRC, on the other hand, does not offer this level of customization and flexibility.

- **Global harmonization**: While DSRC has been adopted in some regions, it has not achieved global harmonization, leading to inconsistencies in spectrum allocation and regulation across different countries. 5G has a more unified approach, with global standardization and broader adoption, making it more attractive for V2X implementations across various regions.

Keeping all of this in mind, automakers have begun to include both in their chipsets. The idea is that cellular networks are the primary communication path, and when those are not available, the chipset will leverage DSRC for peer-to-peer vehicle communication when and where it can.

Connecting to low-powered devices with LoRaWAN

First, let's understand what **Long Range** (**LoRa**) is, as well as its benefits and drawbacks.

LoRa

LoRa operates at Layer 1 (physical) of the OSI model. You could think of it as the equivalent of a Cat-6 RJ-45 cable. You can't do much with that on its own.

> LoRa
>
> LPWAN radio technology was developed by Semtech and designed for use in IoT. It is based on a spread spectrum technique called **Chirp Spread Spectrum** (**CSS**), which allows data to be transmitted over long distances (up to several kilometers) with low power consumption.

Benefits of LoRa

The following are some of the benefits of LoRa:

- **Long range**[14]: 40 kilometers/25 miles (rural environment) and 5 kilometers/3 miles (urban environment)

- **Low power**: Designed to consume minimal energy, with some LoRa-capable devices having built-in batteries that last 20 years. The asynchronous connection model allows the device to sleep when there is no data to transmit or receive.

- **Reliable**: Built-in forward error correction improves resilience against interference.

- **High penetration**: Depending on the region, LoRa operates between 863-928MHz. This frequency range is less than half of the lowest 802.11x band. Due to this, LoRa signals penetrate obstacles approximately twice as well as Wi-Fi.

- **License-free**: See the following table for the unlicensed frequency ranges for LoRa:

---

14   These figures represent best-case line-of-sight range, or the radius of coverage with an omnidirectional antenna. Further, they can vary depending on the design of the gateway.

Region	Band
North America	US915 (902-928MHz)
Europe	EU868 (863-873MHz)
South America	AU915 (915-928MHz)
India	IN865 (865-867MHz)
Asia	AS923 (915-928MHz)

Figure 3.33 – License-free LoRa bands by region

Drawbacks of LoRa

Here are some of the drawbacks of LoRa:

- **Low throughput**: The max bitrate is around 50 kilobits a second

- **Uncontrolled spectrum usage**: License-free operation helps with deployment, but if you have ever had problems with your neighbor's Wi-Fi router stepping on your signal, the potential is apparent.

Long range wide area network (LoRaWAN)

> **Long range wide area network (LoRaWAN)**
>
> This is a protocol that sits on top of LoRA. It operates at Layer 2 (data link) and Layer 3 (network) of the OSI model. LoRaWAN does the same job that Ethernet and IP do for typical computer networks. It is possible to use LoRaWAN on top of a different Layer 1 radio technology, but this is uncommon.

Figure 3.34 – Examples of LoRaWAN gateways

LoRaWAN is an open standard that is supported by the LoRa Alliance, a non-profit organization that promotes the adoption of the technology. It is widely used, having been adopted by many major telcos around the world. LoRaWAN networks are used for applications that require long-range communication, low power consumption, and a low data rate, such as smart metering, asset tracking, and environmental monitoring.

The technology is well suited for non-video IoT applications because it allows rapid deployment of inexpensive sensors and relatively little infrastructure compared, to, say, 5G:

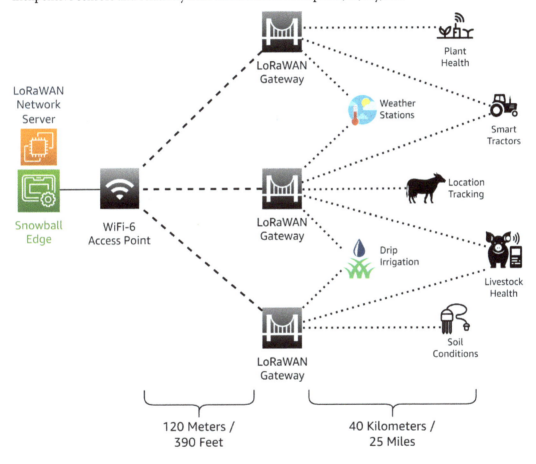

Figure 3.35 – Smart agriculture with LoRaWAN

A LoRaWAN network consists of the following elements:

- **End devices**: These are also called nodes. They are the actual sensors, actuators, cameras, and the like in an IoT deployment. They communicate with gateways over the LoRa protocol.

- **Gateways**: These are also called concentrators. These are similar to Wi-Fi extenders in that they act as a bridge from the end device/node to the network. Unlike Wi-Fi, however, a given device can talk to multiple gateways at once, and all a gateway does is gather those device messages and forward them to the network server. It is up to the network server to handle duplicate messages.

  You usually want your devices to talk to a minimum of three gateways.

  They also have an IP connection of some sort – it could be wired or wireless – so that they can communicate with the network server. That link is not LoRaWAN, because it is an aggregation point and needs higher throughput.

- **Network server**: These could be thought of as similar to the AP controllers some enterprise Wi-Fi networks use to manage multiple access points. They receive messages from the gateways/concentrators and forward them to the application – both over an IP network.

  They are also responsible for deduplication of messages. This is because multiple gateways can receive the same message from a given device, and they will simply forward them along and let the network server figure out if it is unique or not.

  Note that LoRaWAN devices are not paired to a gateway – they are paired to a network server. The gateways are just a transport mechanism.

- **Application server**: This is the final stop of a LoRaWAN message's journey. The application server handles message encryption, data storage, and authentication of new nodes into the network.

LoRaWAN network topology

Notice that, unlike Wi-Fi, LoRaWAN inserts gateways as intermediaries between the devices and the network. While some large enterprise Wi-Fi networks have a similar topology, it is something manufacturers bolted on later for scale-related reasons and is not part of the original 802.11x specification.

With Wi-Fi, a device is only ever associated with one access point at a time, and when they move around, their session is cut over between them. LoRaWAN, on the other hand, sends its traffic to all of the gateways it can see simultaneously. If the server needs to send a message back to the device, it will choose the best gateway to use for that purpose:

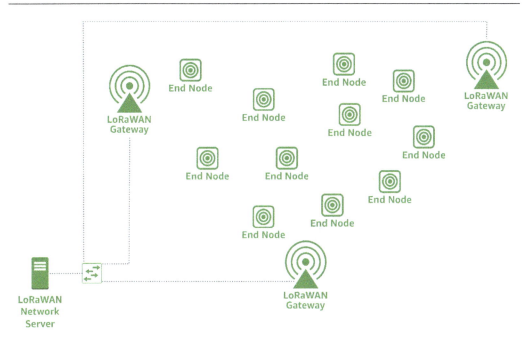

Figure 3.36 – LoRaWAN network topology

This architecture is known as star-on-star. It yields advantages that are relevant to typical LoRaWAN use cases:

- **Redundancy**: If a gateway fails or needs to be taken offline for maintenance, devices in the network are not affected.

- **Affordability**: Because this form of redundancy is a fundamental part of the LoRaWAN specification, it is cheaper to implement both in terms of hardware and deployment effort.

- **Scalability**: The number of gateways a network server can manage is limited only by the processing power of that server. When that is exhausted, additional servers can be added to scale the system horizontally. There are LoRaWAN networks with 40,000 gateways that support many millions of devices [15].

---

15   https://www.thethingsnetwork.org/map

Direct communication between devices

The LoRaWAN protocol does not support direct communication between end nodes. This can be confusing because LoRaWAN-capable devices exist that communicate without involving the gateways. However, this is done using a different protocol such as RadioHead[16] or something proprietary to that manufacturer.

Geolocation

All battery-powered LoRaWAN devices such as tags or sensors can move while they are communicating without increasing the power budget. Additionally, it is not always practical to track physical coordinates when deploying stationary devices.

Fortunately, the LoRaWAN protocol provides two inbuilt methods for determining the position of devices. Nothing needs to be added to existing LoRaWAN-capable endpoints for these to work, making it a lower-cost alternative to adding GNSS to all devices. In some cases, even when the devices do have GNSS positioning, LoRaWAN geolocation is used as a check on that position:

- **Received Signal Strength Indication** (**RSSI**): This measures the received signal power in milliwatts, and is measured in dBm. This method works for coarse positioning in the 1,000-to-2,000-meter range.

- **Time Difference of Arrival** (**TDOA**): Each gateway must have a tightly synchronized time source for this method. Usually, this is obtained from a GNSS network such as GPS. The network server converts the timestamp of when messages were received by each gateway into a distance. It then plots those distances and estimates the devices' location at the intersection. If a device can reach three or more gateways, its position can be calculated to be between 20 and 200 meters.

Regardless of which method is used, a good rule of thumb is that rural deployments will see accuracies toward the lower end of the range while accuracy in urban environments will be toward the higher end. Both methods will benefit from higher gateway density.

LoRaWAN device classes

One of the primary design parameters for LoRaWAN devices is low power consumption. LoRaWAN devices don't leverage any special battery technology. Some of them use simple AAA or AA batteries you can purchase at the supermarket. Rather, it's because they try to spend as much time as possible doing as little as possible.

LoRaWAN device batteries are measured in terms of **milliamp-hours** (**mAh**), just the same as a power bank you might use to recharge your mobile phone. In the LoRaWAN specification, end devices/nodes can operate in three different modes: **Class A**, **Class B**, and **Class C**.

---

16   https://www.airspayce.com/mikem/arduino/RadioHead/

All end devices support **Class A** [14]. These spend most of their time in sleep mode. Because LoRaWAN is not a scheduled protocol, end devices can communicate any time there is a change in a sensor reading or when a local timer on the device goes off:

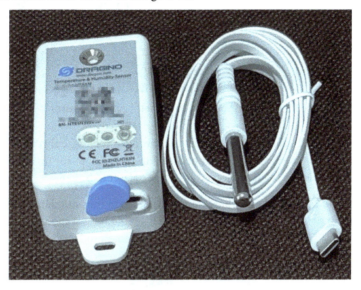

Figure 3.37 – Class A LoRaWAN temperature and humidity sensor

These devices can wake up and talk to the server at any random moment. After the device sends an uplink, it listens for a message from the network one and two seconds after the uplink (receive windows) before going back to sleep. Class A is the most energy efficient and results in the longest battery life. A 5,000mAh power bank for your phone could keep the average class A device running for 30 years [17].

Examples of Class A devices include LoRaWAN-enabled pushbuttons that transmit alarm information in case of an emergency. There are such buttons on the market with a 600mAh capacity that can sustain 70,000 pushes of the button (and associated message transmission).

Class B devices are designed for use in applications where the device needs to transmit data more frequently, but still has relatively low power requirements. They are allowed to transmit data at regular intervals, and they listen for a response from the network after each transmission. This allows them to transmit data more frequently than Class A devices, and the part where they listen for a response ensures more reliability, but they still have a low power consumption:

---

17   Do not attempt this – it is likely such a power bank would self-discharge long before 30 years..

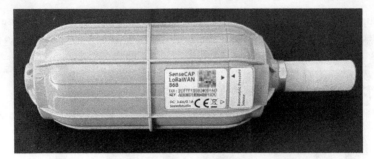

Figure 3.38 – Class B LoRaWAN barometric pressure sensor

Devices in this class might include a smart meter that needs to reliably collect the kilowatt-hour utilization of a power circuit at regular intervals or an environmental sensor that needs to be sure it collects a windspeed sample at prescribed intervals for the dataset to be valid.

Class C devices are used in applications where the device needs to transmit data continuously. They are allowed to transmit data at any time and are always listening for a response from the network. They never go to sleep. This makes them the least power-efficient of the three classes:

Figure 3.39 – Class C LoRaWAN manhole sensor

An example might be a sensor in a manufacturing plant that ensures something dangerous remains within a specific temperature range. Another might be a device that's used for real-time asset tracking, where we want to be actively alerted the moment something leaves the area it is supposed to be in.

Integrating SATCOM

Satellite communication (SATCOM)

SATCOM is the use of satellites to provide communication services, such as telephone, television, and internet connectivity. SATCOM systems use a network of satellites in orbit around the Earth to transmit and receive signals between two or more points on the surface of the Earth, or between the Earth and another body in space (such as a spacecraft).

There are two main types of SATCOM systems: fixed and mobile. Fixed SATCOM systems are typically used to provide communication services to a specific location, such as a remote village or a ship at sea. Mobile SATCOM systems are designed to provide communication services to mobile users, such as aircraft, vehicles, or portable devices.

SATCOM systems are used in a wide range of applications, including military and government communications, emergency and disaster response, and commercial telecommunications. They are particularly useful in areas where it is difficult or impossible to install terrestrial communication infrastructure, such as in remote or inaccessible locations, or disaster-stricken areas.

SATCOM terminal[18]

In the context of satellite communications, a terminal is the user equipment that acts as an interface between the user's network and the satellite constellation. SATCOM terminals vary in cost, size, and complexity, ranging from small handheld devices to larger installations used in industries such as aviation, rail, maritime, and the military. Terminals typically consist of antennas, transceivers, modems, and associated electronics that facilitate satellite communication for voice, data, video, or other forms of communication.

SATCOM frequency bands

For the most part, SATCOM takes place within the SHF or VHF bands, as defined by the ITU. However, SATCOM has its own frequency band definitions, which are more granular:

---

18   Some SATCOM operators refer to terminals as antennas or modems, which is technically inaccurate as a terminal is the overall system the end user needs to connect to.

Band		Frequency (GHz)		Wavelengthn	
Start		Stop	Start	Stop	
Classical L-Band		0.950	1.450	316	207
Extended L-Band		0.950	2.150	316	140
S-band		1.700	3.000	176	100
Extended C-Band	Downlink	3.400	4.200	88	71
	Uplink	5.850	6.725	51	45
LMI C-Band	Downlink	3.700	4.000	81	75
	Uplink	5.725	6.025	52	50
Russian C-Band	Downlink	3.650	4.150	82	72
	Uplink	5.950	6.475	50	46
Standard C-Band	Downlink	3.700	4.200	81	71
	Uplink	5.925	6.425	51	47
X-Band	Downlink	7.250	7.750	41	39
	Uplink	7.900	8.400	38	36
Ku-Band	Downlink	10.000	13.000	30	23
	Uplink	14.000	17.000	21	18
K-Band		18.000	26.500	17	11
Ka-Bandn	Downlink	18.000	21.000	17	14
	Uplink	27.000	31.000	11	10

Figure 3.40 – SATCOM frequency bands

Satellite orbits

Geostationary orbit (GEO)

GEO satellites are positioned in orbit around the Earth at an altitude of about 35,786 kilometers (22,236 miles). They are designed to remain in a fixed location relative to a point on the Earth's surface as they orbit the Earth at the same rate that the Earth rotates.

This makes things easy for ground-based users. There are mobile apps that will tell you exactly where in the sky to point your antenna, and then you're done:

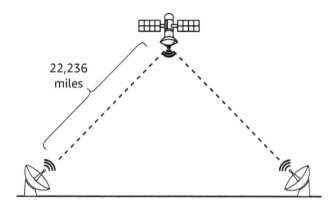

Figure 3.41 – GEO satellite distance

The downside is the high latency incurred when signals have to travel that far. The speed of light is fast, but it is finite. ~200 milliseconds are required for light to go from one spot on the earth up to the GEO satellite and another 200 to go down to another spot. Factor in the latency of any ground segment and a 600ms RTT is considered typical.

Here are some typical GEO-based SATCOM data services:

- **Broadband Global Area Network (BGAN)**: This is an L-band service from Inmarsat. It can achieve speeds up to 492kbps for standard IP data traffic and up to 800kbps for streaming data (usually video), although this depends heavily upon the terminal involved. Six geostationary satellites are involved in providing global coverage (including polar regions) for this service. It is extremely reliable, supporting a 99.9% uptime SLA.

- **Global Xpress (GX)**: This is a Ka-band service from Inmarsat. It can achieve download speeds up to 50mbps and 5mbps speeds for upload. Five geostationary satellites provide near-global coverage.

- **European Aviation Network (EAN)**: This is a hybrid service comprised of a single Inmarsat S-band satellite in geostationary orbit above Europe and Vodafone's terrestrial 4G/LTE network. Specifically built to provide data services onboard aircraft in European airspace, data rates as high as 100mbps are supported. Aircraft use the terrestrial network below 10,000 feet and switch to the S-band service above this altitude.

- **ViaSat-3**: This is a Ka-band service that uses a constellation of three geostationary satellites operated by ViaSat. Each satellite serves a specific region (AMER, EMEA, or APAC), and has a total network capacity greater than 1 terabit per second. Typical consumer plans are 100mbps, while contracts for defense and commercial entities can be higher.

- **GEO HTS**: This is a Ku-band service from SES that can achieve speeds up to 10mbps. It has near-global coverage using four satellites in geostationary orbit.

- **FlexGround**: This is a Ku-band service from Intelsat that supports download speeds up to 10mbps and 3mbps upload speeds. Being one of the pioneers in SATCOM[19], Intelsat has over 50 satellites in geostationary orbit.

> **Low-Earth Orbit (LEO)**
>
> LEO satellites are positioned in orbit around the Earth at an altitude of up to 2,000 kilometers (1,200 miles). Because of this, they are in constant motion relative to an observer.

LEO satellites are known for their ability to provide coverage over a large area of the Earth's surface since they orbit the Earth relatively quickly (compared to GEO satellites). This allows them to provide communication and other services to a large number of users, as well as to track the movement of objects on the surface of the Earth.

The primary technical advantage of LEO-based SATCOM systems is their much lower latency than GEO (as low as ~20ms RTT). The main disadvantage is caused by the fact that they are in constant motion concerning any given point on the ground. They must use mechanisms such as motorized tracking antennas (or complex phased-array antennas) and constellations of a sufficient size to ensure users on the ground can always reach at least one satellite.

Here are some examples of LEO-based SATCOM services:

- **Certus 700**: An L-band service from Iridium that supports speeds as high as 704 Kbps. It is served by 66 cross-linked satellites in LEO.

- **Starlink Roam**: A Ka/Ku-band service from Starlink that supports speeds up to 200 Mbps. It is served by over 3,500[20] cross-linked satellites in LEO, with plans to grow to as many as 12,000.

Global Navigation Satellite System (GNSS)

GNSS is an overarching term that includes all of the systems that use timing signals from satellite constellations to determine a position on the ground for navigation purposes.

---

19   Intelsat launched its first satellite in 1965.

20   As of February, 2023.

GNSS for positioning

> **Trilateration**
>
> All satellite-based navigation systems discussed in this section determine a terminal's position using trilateration. Unlike triangulation, it measures distance – not angles. Satellites in these systems repeatedly broadcast their current position and local time, derived from multiple onboard atomic clocks.

The following figure demonstrates a point on the ground receiving the same broadcast from four satellites:

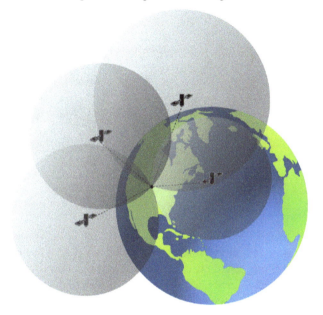

Figure 3.42 – Trilateration using four satellites

From these four pieces of data, a terminal can calculate its position within a margin of error that varies from centimeters to hundreds of meters, depending on the circumstances.

> **Geometric Dilution of Precision (GDOP)**
>
> GDOP is a calculated value that combines the impact of several factors related to the angle at which the ground station can reach the satellites into a single coefficient that expresses how accurate a calculated position is.

Referring back to the previous figure, we can see an example of good geometry of the satellites involved. They are spread across the sky in all three axes. Contrast that with the following situation. In this case, the user is in an area surrounded by mountains. The terminal has no choice but to use samples from satellites that are closer together in the sky, and the calculated position will be less accurate as a result:

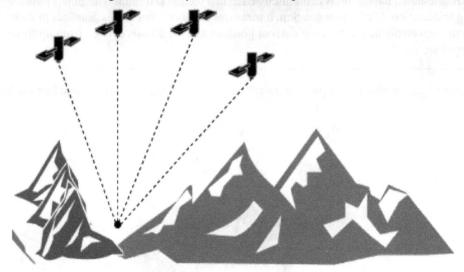

Figure 3.43 – Poor geometry due to obstructions

Other sources of GNSS inaccuracy

Atmospheric refraction is when a satellite's signal is bent a little while traveling through the upper layers of the atmosphere. Sunspot activity can cause interference. Lower-quality receivers are more susceptible to measurement noise, which can happen even under perfect environmental conditions. A clock error of 1 nanosecond (a billionth of a second) can introduce as much as half a meter (1.5 feet) of imprecision.

Urban environments pose a particular challenge to GNSSs. Not only is the geometry compromised by buildings, but the signals the user can receive are often reflected off of them – causing unwanted multipath propagation as previously discussed. If you've ever requested a ride from an app on your phone and wondered why the driver thinks you're at a restaurant two streets away, these are likely culprits.

Global Positioning System (GPS)

The first satellite for what we now know as GPS was launched in 1978 by the United States Air Force. At first, only the US military had access to the system.

In 1983, pilots of a commercial flight from Alaska to Korea made a navigational error that took their aircraft over the Kamchatka Peninsula near Japan. In response, a Soviet SU-15 interceptor shot down the Boeing 747, killing all 269 civilians onboard. To prevent future incidents, the US opened GPS for civilian use.

As of 2020, GPS is operated by the United States Space Force and remains open for anyone to use. At the time of writing, it has 32 satellites in a semi-synchronous[21] **medium Earth orbit** (**MEO**) with an altitude of 20,200 kilometers (12,600 miles). Each orbit has a different inclination, providing global ground coverage.

Global Navigation Satellite System (GLONASS)

Contemporaneously with the rollout of the US's GPS, the Soviet Union began deployment of a similar system known as GLONASS. The first satellite was launched in 1982 and has continued to be developed by the Russian Federation and operated by Roscosmos. Due to economic constraints in the 1990s/2000s followed by sanction-related obstacles in the 2010s, GLONASS has faced numerous challenges. However, it remains operational and available for anyone to use.

Compared to GPS, GLONASS is less accurate on average (though only slightly). That said, due to the different configuration of its orbits, GLONASS is a bit more accurate than GPS at high latitudes (such as within the Arctic or Antarctic circles).

Galileo

Created by the European Union via the European Space Agency, Galileo is a multinational effort to operate a global positioning system that provides independence from single-country control as is seen with GPS and GLONASS. The system went live in 2016 and currently operates 30 satellites in MEO.

At the time of writing, Galileo is the most accurate of the three global systems for the average user.

Regional and augmentation systems

In addition to the three global systems, there are a few regional and augmentation systems. These include the following:

- **Quasi-Zenith Satellite System** (**QZSS**): Operated by Japan, QZSS uses a combination of satellites in geostationary and highly elliptical orbits to augment GPS, improving performance for terminals in Japan and the surrounding region.

- **Navigation Indian Constellation** (**NAVIC**): Deployed by India, NAVIC uses a handful of geostationary satellites to improve performance for GPS terminals in South Asia.

---

21   A semi-synchronous orbit is one in which the spacecraft passes over a given point on the Earth twice per day.

- **Wide Area Augmentation System (WAAS)**: The US **Federal Aviation Agency (FAA)** operates three satellites in geostationary orbit to improve navigation for civilian aircraft in North America.

- **European Geostationary Navigation Overlay Service (EGNOS)**: A distinct system from Galileo, EGNOS is a set of three geostationary satellites that augment GPS for European users. Future plans include the ability to augment the Galileo system as well.

Other uses for GNSS

When a very precise clock source is needed that is accurate down to nanoseconds, expensive atomic clocks are one approach. However, because GNSS satellites have one or more atomic clocks onboard, their signals can be used to indirectly gain access to a free atomic clock. For example, 5G NFV functions, or virtual machines running a **Software-Defined Radio (SDR)** application require access to a physical clock. **Network Time Protocol (NTP)** or **Precision Time Protocol (PTP)** servers frequently save money by making use of GNSS signals.

Summary

In this chapter, we introduced you to elements that are common to all wireless communication technologies that are used at the far edge – concepts such as wavelength, frequency, duplexing, modulation, multipathing, and antenna design.

We built upon that by diving into cellular networking technologies such as 4G/LTE and 5G, reviewing the key advantages of 5G networks and how they enable new low-latency/high-throughput use cases. You were given a survey of LPWAN technologies such as LoRaWAN and NB-IoT, both of which are crucial to use cases such as smart agriculture, V2X, and smart cities.

Finally, we discussed the basics needed to understand SATCOM technologies and the services based on them – upon which the most remote edge computing use cases are dependent.

In the next chapter, we will cover the AWS Snow family of services. These target remote/disconnected edge compute situations.

Part 2: Introducing AWS Edge Computing Services

In *Part Two*, you will be introduced to four key edge computing services offered by **Amazon Web Services** (**AWS**). These are the AWS Snow Family, AWS Outposts, AWS Local Zones, and AWS Wavelength. It explains how these services work in common edge computing scenarios, from remote operations to integrating cloud services with on-site data centers and improving network performance with reduced latency.

This part has the following chapters:

- *Chapter 4, Addressing Disconnected Scenarios with AWS Snow Family*
- *Chapter 5, Incorporating AWS Outposts into Your On-Premise Data Center*
- *Chapter 6, Lowering First-Hop Latency with AWS Local Zones*
- *Chapter 7, Using AWS Wavelength Zones on Public 5G Networks*

4

Addressing Disconnected Scenarios with AWS Snow Family

In today's interconnected world, reliable connectivity is often taken for granted. However, there are numerous scenarios where maintaining a consistent network connection is a challenge, such as remote locations, disaster-stricken areas, or environments with limited or intermittent network access. In these disconnected scenarios, organizations require a solution that can ensure data availability, enable efficient data processing, and one that will support critical operations. This is where the AWS Snow Family comes into play, providing a range of robust and versatile solutions designed specifically to address the unique requirements of disconnected environments.

In this chapter, we will explore how the AWS Snow Family empowers organizations to overcome the limitations of disconnected scenarios and seamlessly bridge the gap between on-premises infrastructure and the cloud. We will delve into the features and capabilities of AWS Snow Family offerings and discuss their use cases, benefits, and considerations. Whether it's securely transferring large amounts of data, performing on-site data processing and analysis, or extending cloud services to the edge, the AWS Snow Family offers reliable, scalable, and cost-effective solutions that cater to the needs of disconnected environments. Join us as we discover the power of AWS Snow to enable data-driven decision-making and unlock new possibilities in disconnected scenarios.

Here are the main headings:

- Introduction to the AWS Snow Family
- Using AWS Snowball Edge
- Using AWS Snowcone

Introduction to the AWS Snow Family

The original AWS Snowball service was introduced in 2015. It started out as a mechanism to move large amounts of data when doing so over the network wasn't reasonable. In the ensuing years, customer demand for new capabilities has driven the expansion of this line into different variants with use-case-specific capabilities:

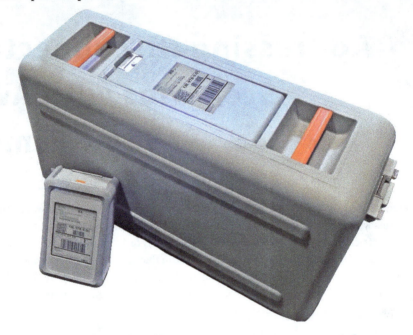

Figure 4.1 – AWS Snow Family devices

All offer an interface and operating model that is consistent with Amazon EC2 and Amazon S3, and they are all designed to run autonomously. All AWS Snow Family devices operate their own local control, management, and data planes. Thus, they do not require a consistent network connection back to the AWS cloud to operate.

AWS Snow Family devices can all host local object storage buckets that utilize the same API/CLI interface as Amazon S3 buckets. When a customer orders one, it is sent to them, they copy their data to these local buckets, and then they ship the unit back to AWS. This is facilitated by an e-ink display on the unit that eliminates the need to pack it in a box or obtain a shipping label separately. When the device is received by AWS, the data is uploaded to the relevant "real version" of the Amazon S3 bucket in question.

Additionally, AWS Snow Family devices do not have the same restrictive environmental requirements as most off-the-shelf compute and storage hardware. AWS Snow Family devices are found operating in a wide variety of field situations that would be impractical with standard off-the-shelf servers. First responders heading to the site of a disaster can even check them in as luggage.

Using AWS Snowball Edge

There is no longer a division between AWS Snowball and AWS Snowball Edge. Now, all such devices fall under the AWS Snowball Edge line, even if their intended use case is a straightforward data migration to S3.

There are four configurations with which an AWS Snowball Edge device can be ordered (see *Figure 4.1*):

	Storage Optimized w/80 TB	Compute Optimized Type 1	Compute Optimized Type 2 1	Compute Optimized w/GPU
HDD in TB	80	39.5	39.5	39.5
SSD in TB	1	7.68	0	7.68
NVME in TB	0	0	28	0
VCPUs	24	52	104	52
VRAM in GB	80	208	416	208
GPU type	None	None	None	NVIDIA V100
10 Gbit RJ45	1	2	2	2
25 Gbit SFP	1	1	1	1
100 Gbit QSFP	1	1	1	1
Volume (in^3)	5381	5381	5381	5381
Weight (lbs)	47	47	47	47
Power draw (avg)	304 w	304 w	304 w	304 w
Power draw (max)	1200 w	1200 w	1200 w	1200 w
Voltage range	100-240 v	100-240 v	100-240 v	100-240 v

Table 4.1 – Comparison of AWS Snowball Edge variants

The AWS Snowball Edge Storage Optimized variant is now used for data migrations in place of the old AWS Snowball. There is a local S3 endpoint to which files can be directly copied using AWS OpsHub, the **AWS Command Line Interface (AWS CLI)**, or direct API commands from a script.

The local compute capacity can be used to host an AWS DataSync instance, an AWS Tape Gateway instance, an AWS File Gateway instance, or another instance that provides a different type of loading interface of your choosing.

---

1   At the time of writing, this variant is limited to US-based regions only

Migrating data to the cloud

Table 4.2 illustrates how long migrations of varying sizes would take depending upon the network throughput:

	50 Mbps	100 Mbps	1 Gbps	2 Gbps	5 Gbps	10 Gbps	25 Gbps	40 Gbps	100 Gbps
50 Terabytes	3.3 months	1.7 months	5 days	2.5 days	1 day	12 hours	5 hours	3 hours	1 hour
500 Terabytes	2.8 years	1.4 years	1.7 months	25 days	10 days	5 days	2 days	1.25 days	12 hours
5 Petabytes	28.5 years	14.3 years	1.4 years	8.5 months	3.4 months	1.7 months	20 days	12 days	5 days
10 Petabytes	57 years	28.5 years	2.8 years	1.4 years	6.8 months	3.4 months	1.3 months	24 days	10 days

Table 4.2 – Comparison of migration times

Many organizations don't have high-throughput internet connections that could be fully dedicated to migration. Nor do they have access to/familiarity with the techniques needed to fully utilize said connection once the latency gets above a few milliseconds.

This is why loading one or more devices connected to a local network and physically shipping to AWS is so popular – despite the days on either end the devices spend on a truck:

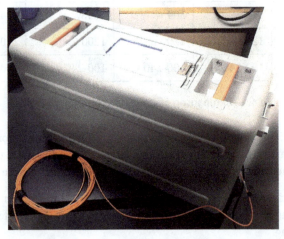

Figure 4.2 – An AWS Snowball Edge device being loaded with data

End-to-end network throughput

Of course, before starting any migration, even to a local device, one must evaluate all of the physical network links involved end to end. Having the AWS Snowball device connected to a 40 GbE switchport via **Quad-Small Form-factor Pluggable** (**QSFP**) won't do much good if an upstream network link operates at a single gigabit:

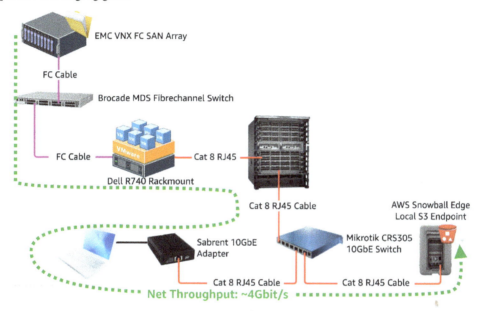

Figure 4.3 – A full end-to-end throughput path

Additionally, there can be choke points on backend **Storage Area Network** (**SAN**) fabrics, disk arrays, **Network-Attached Storage** (**NAS**) devices, or virtualization software somewhere in the middle. In *Figure 4.3*, for example, the data being copied ultimately resides inside **Virtual Machine Disk** (**VMDK**) files on an aging SAN array attached via Fibre Channel (FC) to a server running VMware ESXi.

From the laptop's perspective, the data is being copied over **Common Internet File System** (**CIFS**) from one of the VMware VMs, but in reality, there is a virtualization layer and yet another layer of networking behind that. If, for whatever reason, that SAN array's controller or disk group could only push 4 Gbit/s to the VMware host, it simply doesn't matter that all components of the "normal" network support 10 Gbit/s.

Data loader workstation resources

When transferring data to an AWS Snowball Edge device, it is important to note that the throughput achieved is highly dependent upon the available CPU resources of the machine doing the transfer.

Figure 4.4 – AWS Snowball Edge device loading from a laptop

In *Figure 4.4*, we can see that a reasonably powerful laptop with 8 CPU cores can transfer around 6 Gbit/s, even though there are effectively 10 Gbit/s available end to end on the network. Using a more powerful machine, particularly one with more CPU cores, we would expect the net throughput to rise.

Targets available on AWS Snowball Edge for data loading

There are several types of targets available on an AWS Snowball Edge device that you can use to load data.

NFS endpoint on the AWS Snowball Edge device

This option allows users to access and manage data on the Snowball Edge device using the familiar NFS protocol. This means you can easily mount the Snowball Edge device as a network file share, similar to mounting a NAS device. You can then perform standard file operations such as reading, writing, moving, and deleting files using drag and drop like you would on a departmental file share. Linux or macOS both have NFS support built in, while Windows requires installation of the Services for NFS optional component or a third-party NFS client.

This is generally the most convenient method and the most readily understood. Standard client-side tools such as `rsync`, `xcopy`, Robocopy, or the like can be used with no modifications.

This target has a practical maximum throughput of around 3 Gbit/s.

S3 endpoint on the AWS Snowball Edge device

All members of the AWS Snow Family have a local version of the same sort of S3 endpoint as you would work with in a region. You simply target the S3 endpoint IP on the AWS Snowball Edge device with commands from the AWS CLI or your own code (for instance, a Python script using `boto3`):

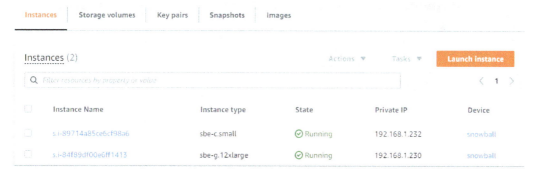

Figure 4.5 – S3 endpoint on an AWS Snowball Edge device

You can also target this local endpoint with third-party programs that know how to work with S3 – common examples include enterprise backup software packages such as Veeam or Commvault.

This target can ingest at speeds in excess of 20 Gbit/s. However, this requires considerable optimization of the client-side transfer mechanism to achieve.

EC2 instance running on the AWS Snowball Edge device

Another approach is to bypass the native endpoints on the device altogether by spinning up an EC2 instance on it:

Figure 4.6 – EC2 instances running on an AWS Snowball Edge device

That instance could run any third-party data transfer software you want, and the limitations on throughput would be specific to that vendor's software.

AWS DataSync agent

The AWS DataSync agent is a special kind of EC2 instance you can spin up on an AWS Snow Family device. It is important to note that this type of target *pulls* the data rather than has data pushed to it like all of the others do. DataSync supports pulling data from the following types of shared storage in your on-premise environment:

- NFS exports

- Windows Server (CIFS/**Server Message Block (SMB)**) shares

- **Hadoop Distributed File System (HDFS)**

- Self-managed object stores (some NAS devices can host S3-compatible stores)

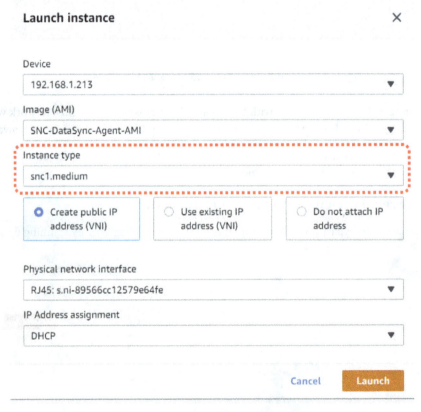

Figure 4.7 – Launching the DataSync agent from OpsHub

You create DataSync tasks inside the AWS Management Console that tell the agent how to access these resources in your environment, when to pull files, how much bandwidth to consume, or if any manipulations need to be done in the process. The agent optimizes the data transfer process by employing techniques such as parallelization, data deduplication, and delta detection to minimize transfer times and optimize bandwidth usage.

A single DataSync task is capable of relaying data to an AWS region at 10 Gbit/s. However, this is dependent upon the resources available within the instance type chosen when the agent is deployed onto the device. At a minimum, an instance type with 2 vCPUs must be used. The more vCPUs the agent has at its disposal, the more it can parallelize the transfer and attain higher speeds.

Client-side mechanisms for loading data onto AWS Snowball Edge

With the exception of AWS DataSync, file loading is a push operation from your data loader workstation. Thus, you will need to use an appropriate client application to communicate with the target you have selected.

> **Performance tip – batching**
>
> Regardless of the client-side mechanism you use to copy data, there is a certain amount of per-file overhead incurred by operations such as encryption. This is why copying a thousand 1 KB files is slower than copying one 1,000 KB file. If the data you are loading consists of many small files spread across many subdirectories, you will probably save time by batching them up into one large archive with utilities such as zip, gzip, or tar. This is true even if you obtain zero compression by doing so.

AWS OpsHub for Snow Family

The simplest thing to do is use the drag-and-drop interface in the AWS OpsHub application. Customers who prefer a GUI interface download and use this anyway to unlock the device and make configuration changes to it. This option also requires no special target configuration.

While it might be convenient, as you can see from *Figure 4.8*, it is also quite slow with a maximum speed of around 0.3 Gbps:

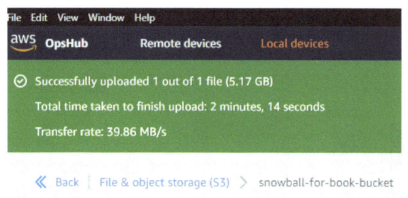

Figure 4.8 – Uploading files via drag and drop in AWS OpsHub

NFS client

When using the NFS endpoint, your data loader workstation must have an NFS client installed. This is usually installed by default on macOS or Linux. While Windows does offer an NFS client, it is not installed by default, and the performance tends to be lower.

AWS CLI

The AWS CLI should be installed anyway on your data loader workstation. It can be used to target the locally running S3 endpoint on the AWS Snowball Edge device. Using the aws s3 sync command, you can do bulk data transfer operations the same way you would with an S3 bucket in an AWS region.

s5cmd

The AWS CLI is a general-purpose utility written in Python. It wasn't explicitly designed to maximize file transfer speed. This means it can't usually push as fast as the S3 endpoint can receive. Fortunately, s5cmd can. It is an open source project available on GitHub. It is written in Go and focuses on maximum parallelization. The more CPU cores your data loader has, the faster it can move data. However, given most laptops or even desktops don't have 128 cores and 25 Gbps interfaces, this option tends to be used when the loader itself is a server in the customer's data center.

Other considerations

Let us assume the following conditions for a migration using an AWS Snowball Edge device:

- A SAN array has two servers as clients
- Each server utilizes two **Logical Unit Numbers** (LUNs) on the SAN
- One server runs Windows Server 2019
- One server runs **Red Hat Enterprise Linux 8** (RHEL 8)
- The Windows server exposes its data for copying through a CIFS share
- The Linux server exposes its data using an NFS export
- The desktop is going to act as a data mover for the AWS Snowball Edge device

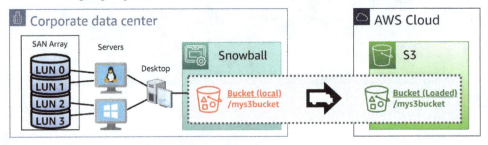

Figure 4.9 – Hypothetical data movement paths

Looking at *Figure 4.9*, we can see several places where the throughput could get slowed down:

- The disk groups/pools on the SAN array
- The controllers/I/O ports on the SAN array
- The Fibrechannel fabric connecting the SAN array to the servers
- The hardware configuration of either server
- The OS and file-serving configuration of either server
- Whether either server is dedicated to this task or is running other apps
- Differences in the CIFS and NFS protocols or their versions
- The network between the servers and the desktop
- Hardware and software configuration of the desktop

An even worse possibility is that the servers and the desktop **can** pull the data from the SAN at maximum speed of all devices and links involved, only to discover this causes the SAN controller to queue I/O requests for a third client you weren't aware of.

It turns out this third server is running a large Microsoft SQL Server that consumes LUNs from the same disk pool on the SAN array, and it also shares the same pair of SAN controllers on the frontend. The 10 Gbit/s of sequential reads causes head thrashing on the disk pool and overruns the shared cache on the controllers.

As a result, the mission-critical application that depends on this database suffers performance degradation – or worse, an outage. Anyone who has overseen many data center migrations – to the cloud or otherwise – has probably witnessed such a situation. Figuring out how fast you can possibly move data onto a device is important, but it is even more important to determine the maximum *non-impactive* speed for the source of the data.

Physical networking

AWS Snowball Edge devices have several Ethernet interfaces you can use to connect them to your network. The interfaces can operate at 1 Gbit/s, 10 Gbit/s, 25 Gbit/s, 40 Gbit/s, or 100 Gbit/s:

Figure 4.10 – Physical network interfaces (PNIs) on AWS Snowball Edge

Interfaces

RJ45: The RJ45 ports on an AWS Snowball Edge device support Ethernet over copper twisted-pair cables at either 1 Gbit/s or 10 Gbit/s. The interface will negotiate one or the other depending on what type of switch port is on the other end. Note that a 10 Gbit/s operation requires, at minimum, a Cat6a cable, or you can expect to drop packets. Cat8 cables are recommended.

Small Form-factor Pluggable (SFP) iteration 28: These are empty slots into which you must insert a transceiver module of some type. You must supply the transceiver module as none ship with an AWS Snowball device of any type. The *28* at the end refers to the fact that they can take Ethernet SFPs that go as fast as 25Gbit/s. These slots are also backward compatible with older 10 Gbit/s or even 1 Gbit/s modules:

Figure 4.11 – 25 GbE fiber optic (left) and 25 GbE RJ45 copper SFPs (right)

With SFP modules, you must supply the correct cable type as well. In the case of the 25 GbE fiber optic SFPs shown in *Figure 4.11*, those would be 50-micron LC-LC OM3 (or better) multimode cables. LC stands for Lucent Connector. They are the smaller squarish connectors that have a receive and transmit strand. OM3 stands for Optical Multimode version 3. These cables typically have an aqua colored jacket, a core size of 50 micrometers. In the case of 25 GbE over copper, a Cat8 twisted pair is required (see *Figure 4.12*):

Figure 4.12 – Cat8 twisted-pair RJ45 cable

Alternatively, 25 GbE SFP28 Twinax cables can be used in these slots. A Twinax cable, also called a **direct-attach copper** (**DAC**) cable, has transceivers on both ends and the cable is molded together as one big unit (see *Figure 4.13*). The cable part inside Twinax is copper, but it isn't twisted-pair. It is essentially two coaxial cables bundled together – hence the name Twinax(ial):

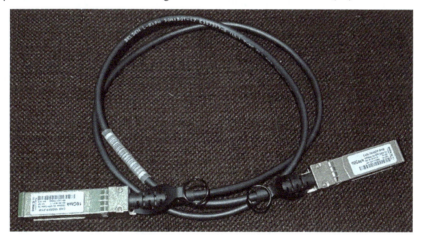

Figure 4.13 – 25 GbE SFP28 Twinax cable

QSFP variant 28 – Like the SFP28 slots, these are empty sockets that you must insert a transceiver into. As is the case with the SFP28 slots, you must supply the transceiver yourself. Whereas SFP28 slots have a single 25 Gbit/s lane, the Quad part of QSFP28 denotes that these have four lanes. They can, therefore, support up to 100 Gbit/s over this single interface. Connectivity options remain the same as with SFP28, but in practice, Twinax cables are almost always used with QSFP. Note that these slots support older 40 Gbit/s modules as well:

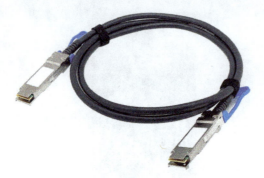

Figure 4.14 – 100 GbE QSFP28 Twinax cable

Logical networking

First, we must level set on some terms that have specific meanings within the context of AWS Snowball Edge. These terms differ a bit from what you see in an EC2 VPC:

- **Public IP**: In this context, the term "public" does not mean a routable IP on the internet. It simply means an IP address on the "outside" network of the device – this is the network your device will acquire **Dynamic Host Configuration Protocol (DHCP)** addresses from when you plug it in for the first time. The default gateway for this network will be a router that you own. DNS and NTP will also be pointing toward addresses you use now on your network.

- **Private IP**: An IP address on the "inside" network of the device. Perhaps confusingly, AWS has chosen to make the private network range on all AWS Snow Family devices `34.223.14.128/25`. This cannot be changed, and yes – it is a routable prefix registered to AWS with the **Internet Assigned Numbers Authority (IANA)**. There are no services attached to the "real" version of this prefix out on the internet, so don't worry.

- **Virtual Network Interface (VNI)**: A static 1:1 NAT mapping of a public IP to a private IP. These are needed for EC2 instances to talk to any network outside of the private range inside the device.

- **Direct Network Interface (DNI)**: This is a way to map one of the physical RJ45/10 GbE Ethernet ports on the AWS Snowcone device to an EC2 instance inside the device, thus bypassing the 1:1 NAT translation from `34.223.14.128/25` to `192.168.x.x` (or whatever your network's IP range is).

Two VNIs each on a different physical Ethernet port

Configuring an AWS Snowball Edge device with two VNIs, each on a separate physical Ethernet port, offers several key benefits. First, it provides increased network bandwidth and throughput by leveraging the capabilities of two separate network connections. This is particularly advantageous in scenarios that require high-speed data transfer or processing, allowing for faster and more efficient operations.

Secondly, having separate physical Ethernet ports for each VNI allows for network segregation and isolation at a hardware level. This enables the Snowball Edge device to maintain a strict separation between different types of network traffic or data flows. By keeping the networks isolated, organizations can ensure enhanced security, compliance, and operational control over their data and applications.

Furthermore, the configuration with two separate physical Ethernet ports provides inherent redundancy and **high availability** (**HA**). If one network connection or port experiences an issue, the Snowball Edge device can automatically switch to the other port, maintaining uninterrupted connectivity and data transfer. This redundancy ensures continuity of operations and minimizes the impact of any network failures:

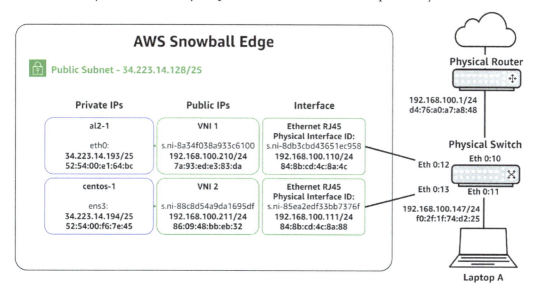

Figure 4.15 – AWS Snowball Edge device with two VNIs on separate PNIs

AWS Snow "private" subnet – 34.223.14.128/25

Something you will notice right away in *Figure 4.15* is that the EC2 instances are configured for an internal network of 34.223.14.128/25, which is a routable prefix on the internet. At the same time, the "public" IPs mapped to them on their VNIs live on 192.168.100.0/24 – a non-routable *RFC 1918* address space. This is counter-intuitive and the opposite of how public subnets work inside an AWS region.

Rest assured this is done for a reason. AWS owns the 34.223.13.128/25 space with IANA, and it is not actually used on the internet. AWS chose to do this to make deployment of Snow Family devices simpler by ensuring the default private subnet is never the same as whatever *RFC 1918* address space a customer is using.

Note that while you can make the "public" subnet of the VNIs live on whatever you wish, it is not possible to change the "private" subnet on any AWS Snow Family device – it is always 34.223.14.128/25.

Two VNIs sharing one physical Ethernet port

In certain situations, physical constraints may prevent the ideal configuration of separating VNIs onto different Ethernet ports on an AWS Snowball Edge device:

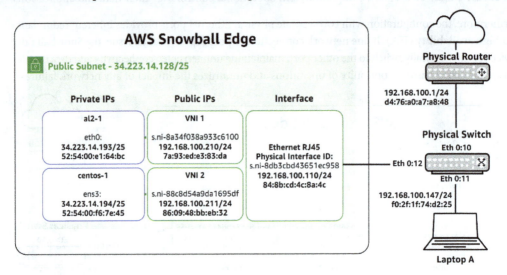

Figure 4.16 – AWS Snowball Edge device with two VNIs on a single PNI

Figure 4.17 illustrates the two network paths possible from EC2 instance al2-1. This instance can connect to devices outside the AWS Snowball Edge environment via VNI 1, which is configured as a 1-1 NAT entry for 192.168.100.210 mapped to the IP the EC2 instance has configured internally of 34.223.13.193:

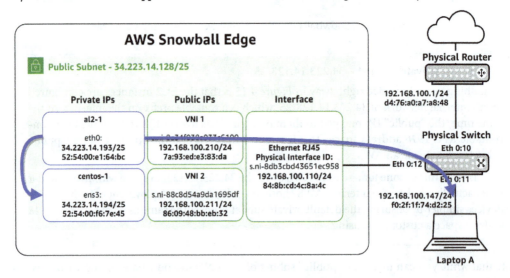

Figure 4.17 – AWS Snowball Edge network flows with VNIs

At the same time, `al2-1` can communicate directly to `centos-1` across the AWS Snowball Edge device's internal subnet of 34.223.14.128/25.

> **VLANs on AWS Snow Family**
>
> It is possible for a Snow Family device to have VNIs that share the same physical Ethernet port configured for two different *RFC 1918* subnets through the use of VLAN tagging. This helps to mitigate some security concerns expressed by customers, but be aware: instances will always be able to talk directly on the internal 34.223.14.128/25 subnet. It is therefore important that security groups are used to limit this.

DNIs

DNIs were introduced to AWS Snow Family devices to support advanced network use cases. DNIs provide layer 2 network access without any translation or filtering, enabling features such as multicast streams, transitive routing, and load balancing. This direct access enhances network performance and allows for customized network configurations.

DNIs support VLAN tags, enabling network segmentation and isolation within the Snow Family device. Additionally, the MAC address can be customized for each DNI, providing further flexibility in network configuration:

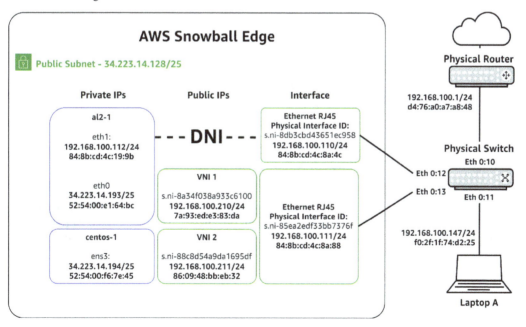

Figure 4.18 – AWS Snowball Edge device with one DNI

> **DNIs and security groups**
>
> It's important to note that traffic on DNIs is not protected by security groups, so additional security measures need to be implemented at the application or network level.

Snowball Edge devices support DNIs on all types of physical Ethernet ports, with each port capable of accommodating up to seven DNIs. For example, RJ45 port *#1* can have seven DNIs, with four DNIs mapped to one EC2 instance and three DNIs mapped to another instance. RJ45 port *#2* could simultaneously accommodate an additional seven DNIs for other EC2 instances.

Note that the Storage Optimized variant of AWS Snowball Edge does not support DNIs:

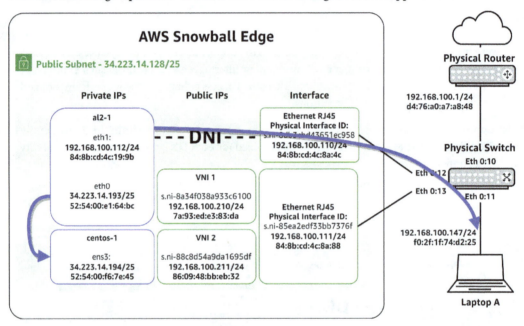

Figure 4.19 – AWS Snowball Edge network flows with DNIs

Looking at *Figure 4.19*, we can see that `al2-1` has two Ethernet ports configured inside Linux. One is on the typical 34.223.14.128/25 subnet, but the other is directly on the 192.168.100.0/24 RFC 1918 space. A configuration such as this is the only time an interface on an EC2 instance on an AWS Snow Family device should be configured for any subnet but 34.223.14.128/25.

Figure 4.20 shows what a DNI looks like from the perspective of the EC2 instance that has one attached:

Ethernet interface
/0/100/4

product: X553 Virtual Function [8086:15C5]
vendor: Intel Corporation [8086]
bus info: pci@0000:00:04.0
logical name: eth1
version: 00
serial: ba:69:5f:b4:1b:53
size: 10Gbit/s
capacity: 10Gbit/s
width: 64 bits
clock: 33MHz
capabilities:
  MSI-X,
  PCI Express,
  bus mastering,
  PCI capabilities listing,
  ethernet,
  Physical interface,
  10Gbit/s (full duplex)

configuration:
  autonegotiation: off
  broadcast: yes
  driver: ixgbevf
  driverversion: 4.1.0-k
  duplex: full
  ip: 192.168.1.237
  latency: 0
  link: yes
  multicast: yes
  speed: 10Gbit/s
resources:
  irq: 0
  memory: febc0000-febc3fff
  memory: febc4000-febc7fff

Figure 4.20 – DNI details under Amazon Linux 2

Storage allocation

All AWS Snowball Edge device variants work the same way with respect to storage allocation. Object or file storage can draw from the device's HDD storage capacity, while block volumes used by EC2 instances can be drawn from either the device's HDD or SDD capacity. *Figure 4.21* shows an example of this:

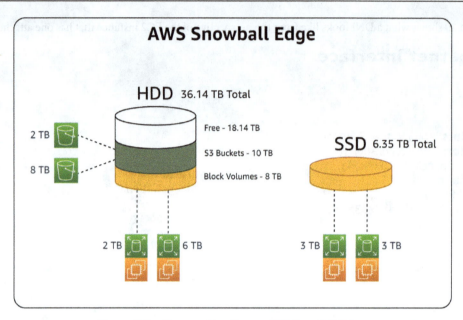

Figure 4.21 – Storage allocation on AWS Snowball Edge

S3 buckets on a device can be thought of as being thin-provisioned in the sense that they start out consuming 0 bytes, and as objects are added, they only take the amount needed for those objects from the HDD capacity.

Block volumes for EC2 instances, on the other hand, can be thought of as thick-provisioned. When the volume is created, a capacity is specified, and it is immediately removed from the HDD pool for any other use.

Using AWS Snowcone

AWS Snowball and AWS Snowball Edge are physically large devices. They both weigh around 50 lbs (~23 kg) and are comparable in size to a suitcase. While functionally similar to the AWS Snowball Edge service, AWS Snowcone comes in a much smaller form factor. AWS Snowcone is best suited for migrations of smaller amounts of data, or situations where a small amount of compute is needed to act as an IoT gateway.

Table 4.3 compares the two primary variants of AWS Snowcone devices available to order.

	HDD Version	SSD Version
HDD in TB	8	0
SSD in TB	0	14
NVME in TB	0	0

	HDD Version	SSD Version
VCPUs	2	2
VRAM in GB	4	4
GPU type	None	None
10 Gbit RJ45	2	2
25 Gbit SFP	0	0
100 Gbit QSFP	0	0
Wi-Fi generation	Wi-Fi 5	Wi-Fi 5
Wi-Fi MIMO	2x2	2x2
Volume (in3)	170	170
Weight (lbs)	4.5 lbs	4.5 lbs
Power draw (avg)	25 w	25 w
Power draw (max)	45 w	45 w
Voltage range	100-240 v	100-240 v

Table 4.3 – Comparison of AWS Snowcone variants

In general, AWS Snowcone works the same way as an AWS Snowball Edge device. For example, you still use the same utilities to manage them, such as AWS OpsHub or the AWS Snowball Edge client. However, there are some noteworthy exceptions.

Wi-Fi

Something you'll notice in *Table 4.3* is the availability of Wi-Fi[2] on AWS Snowcone – something not found on its larger cousins. For maximum throughput, it is recommended to use a Wi-Fi **access point** (**AP**) that has two antennas and supports `802.11ac` with 2x2 MU-MIMO in 5 GHz mode.

Power supply

The second difference to take into account is the lack of internal power supply on AWS Snowcone devices. You will need to provide a power adapter that connects via USB-C to the device and to a wall outlet relevant to your region. Alternatively, it is possible to use a USB-C battery for more mobile use cases. In either case, you must ensure the adapter chosen is rated for 45 watts or greater.

2   Wi-Fi is available only with AWS Snowcone devices ordered from North American regions

Data transfer targets

Both the AWS DataSync agent and NFS endpoints are available on AWS Snowcone, and they work the same way as they do on an AWS Snowball Edge device. However, while you do still load a local replica of an S3 bucket in the cloud, the local S3 endpoint is not available. This means you cannot use the AWS CLI or s5cmd clients on your data loader workstation.

Summary

In this chapter, we reviewed how the AWS Snow Family presents a comprehensive suite of solutions that empower organizations to effectively address disconnected scenarios and overcome the challenges of limited or intermittent network connectivity. We talked about how businesses can securely transfer, store, and process large amounts of data in remote or offline environments. We covered how the robustness and versatility of these devices enable organizations to extend their cloud infrastructure to the edge, perform data analysis and processing at the point of collection, and ensure data availability in challenging circumstances.

With AWS Snow, the digital transformation journey extends beyond the boundaries of traditional network connectivity, providing a pathway to success in even the most challenging and remote environments.

In the next chapter, we will explore incorporating AWS Outposts into your on-premises data center.

5

Incorporating AWS Outposts into Your On-Premises Data Center

The original AWS Outposts service was introduced in 2019. It is best thought of as an extension of an existing AWS region into an on-premises data center.

AWS Outposts was developed due to customers' desire to expand their use of AWS' cloud-based services while operating under constraints associated with latency or regulatory/compliance regimes that dictate the physical location of compute/storage hardware. Compared to the AWS Snow family of services, AWS Outposts offers considerably more compute and storage capacity, an expanded list of supported services, and a user experience more closely matching that found within standard AWS regions.

We will cover the following topics in this chapter:

- Introducing AWS Outposts
- Using AWS Outposts Rack
- Using AWS Outposts Server

Introducing AWS Outposts

Unlike the AWS Snow family of services, AWS Outposts require permanent, reliable network connectivity back to a parent region. This is because the control and management plane depend on the parent region to function properly. Another key difference is that the hardware involved is more like that seen in your on-premises data centers today. Similar environmental requirements to standard rackmount servers apply, unlike the highly ruggedized self-contained devices seen in the Snow family.

AWS Outposts provides fully managed compute, storage, and database services in a hybrid cloud deployment. A subset of these capabilities is offered locally for low-latency connectivity and local data processing, with the remainder offered via a dedicated link back to an AWS region. It is offered in two form factors – 1U or 2U Outposts servers and 42U Outposts racks.

An AWS Outposts rack deployment consists of up to 96 42U racks in your on-premises data center. While you can order from a selection of physical server types and storage configurations, AWS handles the physical installation and configuration within each rack. All of the racks in a given deployment constitute a single logical entity and can share networking and storage resources among any instance in the deployment.

AWS Outposts servers, on the other hand, are self-contained islands of compute capacity. While it is possible to have multiple AWS Outposts servers in the same on-premises data center, from a logical perspective, they are separate deployments with their own connections back to the parent region. They cannot form a cluster or share resources.

While the physical servers found in both offerings appear similar to off-the-shelf models you've worked with before, they are custom-manufactured for AWS. They are the same physical servers you would find hosting EC2 instances inside an AWS region. This affords several advantages, all of which will be explored later in this chapter (see *Using AWS Outposts servers*).

Using AWS Outposts Rack

First, let's define some terms that represent concepts that carry specific meanings within the context of AWS Outposts rack:

- **AWS Outposts Rack**: A managed service provided by AWS that is distinct from AWS Outposts server. One deployment of AWS Outposts rack can consist of multiple physical racks.

- **Rack**: When the word rack is used without the AWS Outposts prefix, it refers to a specific 42U 19" preconfigured unit provided by AWS. Each rack houses some combination of compute, storage, and/or networking hardware.

- **Deployment**: One or more racks, all of which operate in concert and share certain logical constructs, such as a service link or **Local Gateway** (**LGW**). A deployment can consist of up to 96 racks and is logically tied to a specific **Availability Zone** (**AZ**) within its parent region.

- **Site**: The physical location where one or more deployments reside.

- **Resilience zone**: A concept within an Outposts rack that represents a failure boundary for resources, enhancing fault tolerance and availability. A common use of these is to ensure two EC2 instances run within different racks of a deployment.

Use cases

Here are some common customer use cases for all members of the AWS Outposts family:

- The need for very low latency and high-throughput connectivity to AWS services from customers' on-premises infrastructure

- The need for analytical tools used for ML training, forecasting, and more to be positioned close to large amounts of data that are unable to move to the cloud because of bandwidth constraints, time constraints, or cost concerns

- The need to leverage local AWS capabilities from legacy applications that are poor candidates for migration to the cloud because they contain low-latency integrations with other on-premises systems

- Data sovereignty laws that require that its citizens' data only be accessed from within that country's borders, making it necessary to bring the needed cloud capabilities into the country applying those regulations

To contrast AWS Outposts rack with on-premises installations, we will walk through the ordering process and the elements you must consider when planning a deployment.

Ordering an AWS Outposts rack

Customers can choose options such as storage such as Amazon **Elastic Block Store** (**EBS**) gp2 volumes and Amazon **Simple Storage Service** (**S3**), compute options such as EC2 instance type configurations, or other services that might impact the physical configuration, such as **Relational Database Services** (**RDS**).

The payment options include fully upfront, partially upfront, and monthly – all of which include a 3-year commitment:

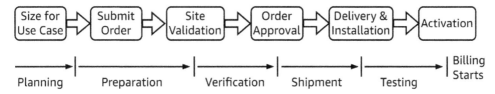

Figure 5.1 – The ordering process for AWS Outposts Rack

The process of provisioning an Outposts rack is unlike provisioning most other AWS services. First, customers work with AWS personnel to identify use cases and capacity requirements. Using these requirements, customers order specific SKUs to be delivered at defined customer locations. AWS personnel visit these locations to conduct site surveys. This is done to verify power availability, HVAC capacity, air quality, whether seismic tie-downs are needed, load capacity for raised floors, and other such considerations. Customers meet the AWS white-glove installation team at a designated time for delivery, installation, and instantiation of Outposts. Next, tests to verify the failover of redundant components and control plane checks are run. Only after everything is validated is management handed over to the customer. 24 hours later, billing begins.

Physical elements

The physical design of an AWS Outposts rack reflects a meticulous blend of cutting-edge technology and ergonomic engineering, encapsulating the essence of a compact, self-contained cloud infrastructure that seamlessly extends the AWS ecosystem to on-premises environments. Designed for optimal efficiency and scalability, an AWS Outposts rack is a refined integration of compute, storage, and networking components, elegantly packaged within a standard 19-inch rack form factor.

AC power requirements

Each AWS Outposts rack in a given site supports three power configurations – 5 kVA (4 kW), 10 kVA (9 kW), or 15 kVA (13 kW). This is because it is not uncommon that a given data center limits the amount of power available to a rack or cage. Single-phase and three-phase redundant AC options are available for all three options. Check the *Requirements* section of the *User Guide for racks* for more details: `https://docs.aws.amazon.com/outposts/latest/userguide/outposts-requirements.html`.

DC power bus

While the incoming power from your data center to an AWS Outposts rack is standard AC, within each rack, the servers are all supplied via a DC power bus. This differs from standard AC **Power Distribution Units (PDUs)** found inside the racks of most on-premises data centers.

In a standard AC PDU, AC power is converted into DC power using power supplies within the servers. With a DC power bus, this conversion is done centrally. Physical servers in an AWS Outposts rack do not need individual power supplies. This increases the efficiency of power utilization, thereby improving the cost-effectiveness and sustainability parameters of the physical footprint.

Don't worry about the specifics; each rack in a deployment will arrive preconfigured appropriately. If you ever order additional servers to expand a rack, an AWS installer will handle this cabling and all of the calculations involved.

Outpost networking devices (ONDs) and customer networking devices (CNDs)

Each rack has two top-of-rack switches installed; these are referred to as ONDs. Two switches that meet the requirements outlined in the `User Guide for racks` are designated as CNDs.

At the physical network layer, a leaf/spine topology is used between all racks in a deployment. Each OND is interconnected to all others in a deployment to form the leaves. Further, each rack is connected to both CNDs, which act as the spine.

It is recommended that each OND within a given rack is connected to a separate CND via a top-of-rack **Patch Panel** (**PP**) to provide network diversity for both logical and physical connections:

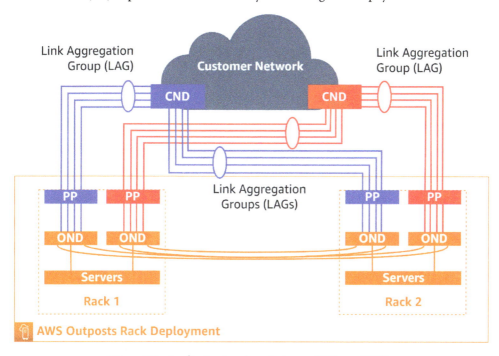

Figure 5.2 – Leaf/spine topology between ONDs and CNDs

It is further recommended that each of these links be more than one physical cable configured in a **Link Aggregation Group** (**LAG**), also known as an IEEE 802.1q Ethernet trunk. Note that even if only one cable is used for a given link, a LAG for it is still provisioned to allow the customer to add additional links later more easily. The preceding figure illustrates this concept.

Customers can work with network hardware vendors that offer specific solutions for AWS Outposts Rack. This helps smooth the deployment with proven configurations. One example is Juniper Networks. Juniper provides documentation that walks customers through configuring their QFX Series devices to interoperate with AWS Outposts. Juniper suggests leveraging their QFX5000 Series in a top-of-rack position and either the QFX5000 or the QFX10000 Series in the spine role for use with AWS Outposts. Using this configuration will enable customers to deploy data center fabric capable of providing 400G connectivity while providing 100G connectivity to single Outposts racks. The following figure depicts the suggested integration of Juniper QFX Series devices and AWS Outposts racks:

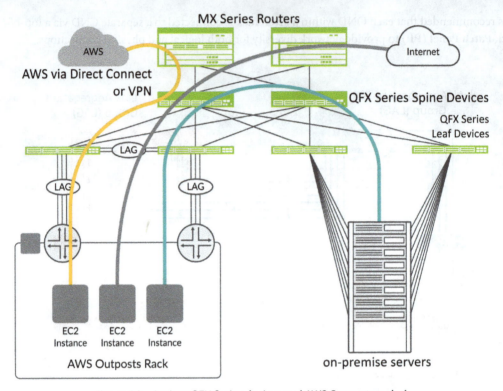

Figure 5.3 – Juniper QFX Series devices and AWS Outposts racks [1]

The orange path in the preceding figure shows the path EC2 instances in an AWS Outposts rack take to natively access AWS resources. The gray path shows how EC2 instances can access the internet or enterprise WAN. Finally, the green path shows the way EC2 instances can reach on-premises resources locally.

Maintenance

When an Outposts rack requires physical maintenance or hardware replacement, AWS works with the customer to schedule AWS personnel to come to the facility and replace the hardware. Customers can use the provided Nitro Security Key to cryptographically shred data according to NIST standards before returning it to AWS.

Logical networking

Now that we've covered the physical networking elements, let's shift focus to the logical layer. This is where we will introduce key AWS networking concepts unique to AWS Outposts Rack.

---

1   Image Credit Juniper Networks

Extending an AWS region

AWS Outposts racks offer customers the ability to seamlessly extend their existing Amazon **Virtual Private Clouds** (**VPCs**) into their AWS Outposts deployment. From the perspective of an EC2 instance in a subnet on AWS Outposts, things are no different than if it had been launched in-region. It can route to instances within the same VPC just like two instances in different AZs would. Constructs defined on a per-VPC basis such as security groups, routing tables, or DHCP options can be leveraged normally. In this sense, an Outposts rack deployment can be thought of as a little piece of an AWS region that only you have access to.

Service link

Imagine taking a rack from an AWS data center and installing it in your on-premises data center. You would need to somehow allow the compute and storage components within the rack to communicate with the management and control plane elements within the region you took it from. Because it is not practical to run a pair of really long Ethernet cables, something else must perform this function. This is something AWS calls a service link:

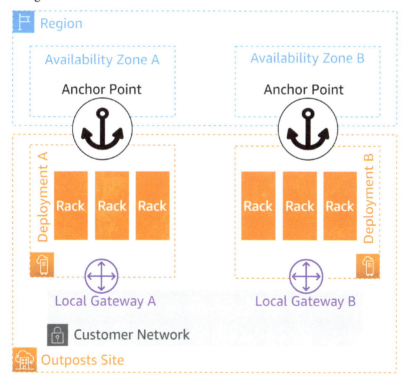

Figure 5.4 – Service link mapping to an AZ in the parent region

There is only one service link per deployment in a logical sense, although redundancy at the physical layer is crucial. The throughput required for a given deployment's service link is determined by many factors, some of which include the following:

- Sizes of **Amazon Machine Images (AMI)** used

- Amount of VPC traffic to the AWS region needed

- Whether the customer needs to size for peak or average utilization

- Number of racks in the deployment

- Configuration of the compute capacity/physical servers within those racks

The service link is implemented as a series of VPN tunnels that segment the intra-VPC traffic from the management traffic between the AWS region and the Outpost. Ideally, these tunnels are mapped to separate VLANs at layer 2 to simplify troubleshooting, but this is not required.

Customers can create specific routes in their service link VLAN or use the quad-zero (0.0.0.0/0) route via one of two options. First, we have AWS Direct Connect's public **virtual interface (VIF)**, which terminates on the AWS region's public IP space:

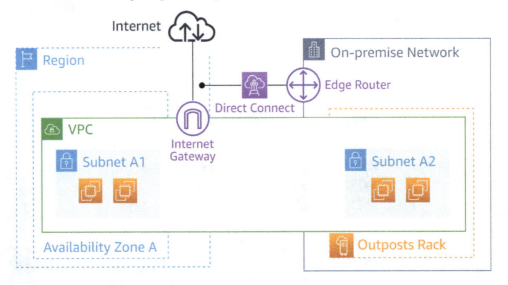

Figure 5.5 – Service link utilizing an AWS Direct Connect public VIF

Then, we have the **over the public internet** approach. It is straightforward because the service link VPN tunnels are instantiated outbound from the AWS Outposts rack deployment to the associated region:

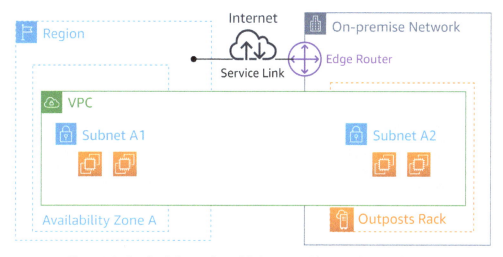

Figure 5.6 – Service link over the public internet with no AWS Direct Connect

The following table shows what ports are required for the service link:

Protocol	Source Port	Source Address	Destination Port	Destination
UDP	443	Outpost service link /26	443	Parent region's public IP space
TCP	1025-65535	Outpost service link /26	443	Parent region's public IP space

Figure 5.7 – Ports required for the service link

Once this connection has been established, all traffic between the associated VPCs and the AWS Outposts Rack deployment flows through it. This includes such things as customer data plane traffic, as well as all management traffic, such as firmware/software updates, control plane traffic, and internal resource monitoring flows. The only traffic that does not go over the service link is that which is bound for the LGW.

LGW

A common reason customers deploy AWS Outposts is because they have something in their on-premises data center, such as a database, and the application has very tight latency or jitter requirements. It wouldn't do much good if an EC2 instance on an AWS Outposts instance had to route back over the service link to get to an internet gateway or NAT gateway just to turn around and go back to the same data center:

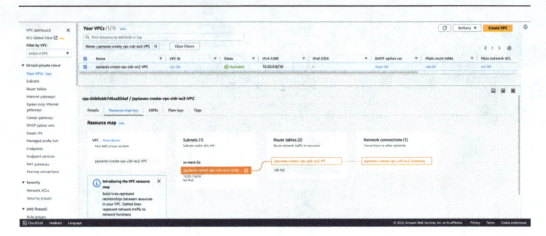

Figure 5.8 – Outposts subnet routing to the internet via an internet gateway

Alternatively, if an EC2 instance on an AWS Outposts Rack deployment wishes to connect to the internet, it could do so by traversing the service link back to the region to utilize the internet gateway for the VPC:

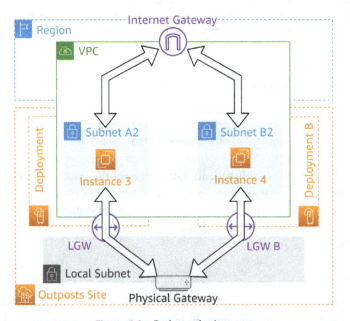

Figure 5.9 – Paths to the internet

However, an LGW provides a more direct path for such instances to leave the VPC. LGWs are also what an instance would use to connect to a resource in the on-premises data center, such as the aforementioned database.

Elastic load balancers in AWS Outposts Rack deployments

An **Application Load Balancer** (**ALB**) can be used to distribute incoming HTTP(S) traffic to different types of resources on Outposts racks, such as EC2 instances and containers.

Available services

An AWS Outposts Rack deployment provides several locally available services that are manageable via the same APIs or **Infrastructure as Code** (**IAC**) tools, such as CloudFormation.

Amazon Route 53 Resolver

This service allows customers to provide on-premises resources with local DNS resolution. Outposts racks can use this capability to provide hybrid DNS functionality between an on-premises DNS server and a Route 53 Resolver through the use of inbound and outbound endpoints. This also provides customers with the ability to cache all DNS queries originating from the Outposts rack for improved application availability by eliminating additional DNS requests to be sent after the cached entry is made:

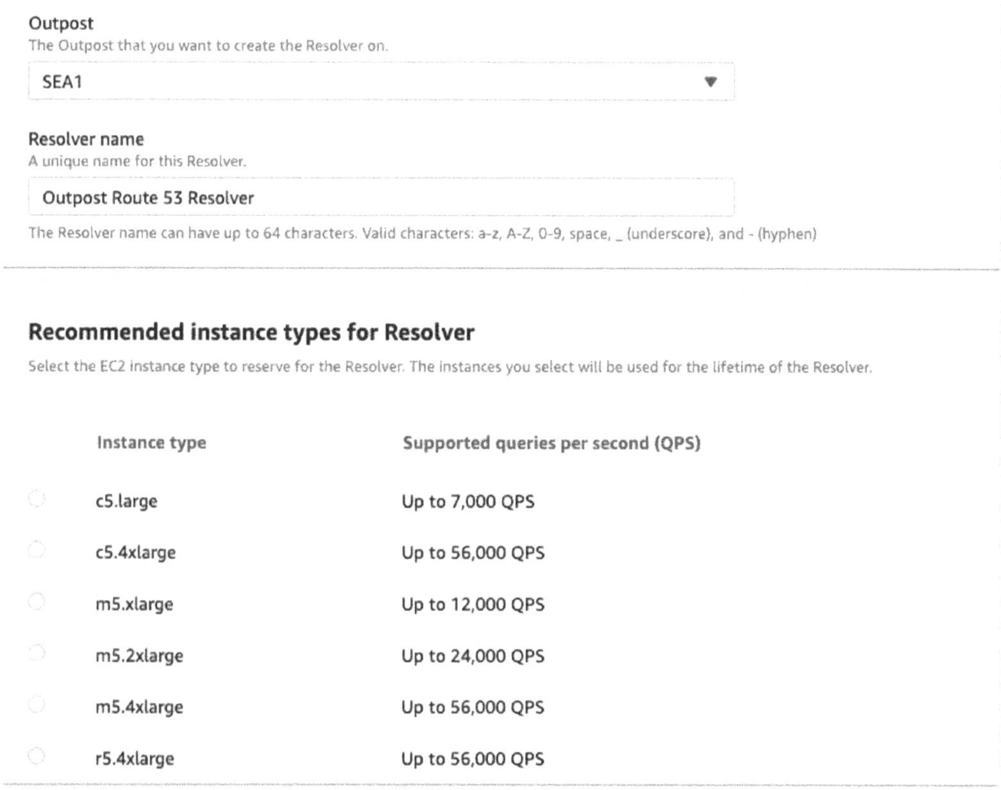

Outpost
The Outpost that you want to create the Resolver on.

SEA1 ▼

Resolver name
A unique name for this Resolver.

Outpost Route 53 Resolver

The Resolver name can have up to 64 characters. Valid characters: a-z, A-Z, 0-9, space, _ (underscore), and - (hyphen)

Recommended instance types for Resolver

Select the EC2 instance type to reserve for the Resolver. The instances you select will be used for the lifetime of the Resolver.

Instance type	Supported queries per second (QPS)
c5.large	Up to 7,000 QPS
c5.4xlarge	Up to 56,000 QPS
m5.xlarge	Up to 12,000 QPS
m5.2xlarge	Up to 24,000 QPS
m5.4xlarge	Up to 56,000 QPS
r5.4xlarge	Up to 56,000 QPS

Figure 5.10 – Creation of an Amazon Route 53 Resolver

Note that the following capabilities will not be available during any network disconnect event that occurs between the deployment and parent region: health checks, control plane changes, DNS failover, cached entry retries which can result in stale responses, and of course connectivity to in-region VPC resources, even though DNS resolution for those resources will continue to function properly.

Amazon EC2

AWS Outposts Rack deployments can host the following instance types:

- M5/M5d for general-purpose workloads
- C5/C5d for compute-intensive workloads
- R5/R5d for memory-optimized workloads
- G4dn for graphics-optimized workloads
- I3en for I/O optimized workloads:

Figure 5.11 – Deploying an EC2 instance to a subnet within an Outposts deployment

AWS has plans to begin offering EC2 VT1 instances for better video transcoding capabilities and AWS Arm-based Graviton powered instances such as the C6g for compute-intensive workloads, M6g for general-purpose workloads, and the R6g for memory-intensive workloads for AWS Outposts racks soon.

For higher availability, EC2 instances in a deployment can utilize placement groups to provide fine-grain control over instance locations helping to reduce hardware failure risks.

Amazon EBS

EBS is an easy-to-use, scalable, high-performance block-storage service designed for EC2. Along with local instance storage, an AWS Outposts Rack deployment offers gp2 EBS volumes for general-purpose block storage needs:

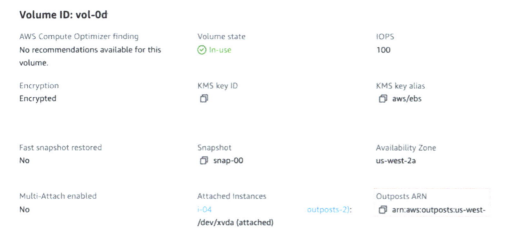

Figure 5.12 – An EBS volume in an AWS Outposts Rack deployment

When ordering, there are three specific tiers of EBS storage available in a given rack: 11 TB, 33 TB, and 55 TB.

Amazon S3

S3 is an object storage service that offers industry-leading scalability, data availability, security, and performance. Customers of all sizes and industries use Amazon S3 to store and protect data for a range of use cases, such as data lakes, websites, mobile applications, backup and restore, archive, enterprise applications, IoT devices, and big data analytics.

S3 storage in an AWS Outposts Rack deployment is provided by the purpose-built OUTPOSTS S3 storage class. The S3 data is stored across multiple devices and servers and accessed with API calls over access points and endpoints within a VPC:

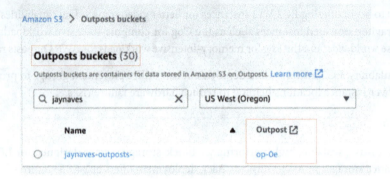

Figure 5.13 – An S3 bucket on an AWS Outposts Rack deployment

When ordering, a given rack in a deployment can be provisioned with 26 TB, 48 TB, 96 TB, 240 TB, or 380 TB of S3 storage capacity. Customers can also add S3 storage capacity to an existing Outposts deployment down the road.

Within Outposts, S3 bucket names have to be unique across the Outpost and AWS account requiring the bucket name, account ID, and outpost ID to be identified. The following is an example that shows the required appending of "-outposts" to the S3 service and outpost/outpost ID as the resource type:

arn:aws:s3-outposts:region:account-id:outpost/outpost-id/bucket/bucket-name

Using S3 buckets in a deployment can serve as the storage location of EBS volume snapshots taken within that deployment. This ensures that data stays within a customer's on-premises location. This is critical for use cases involving data residency requirements.

Amazon RDS

RDS is a managed service that makes it simple to set up, operate, and scale databases in the cloud. The following database engines are available within an AWS Outposts Rack deployment:

- RDS for SQL Server
- RDS for PostgreSQL
- RDS for MySQL

These databases have the advantage of being close to on-premises applications that require low latency or local connectivity. The connectivity to your AWS region is required for backups, restores, snapshots, and transaction logs. Note that customers can restore a snapshot from one AWS Outposts Rack deployment to another via the associated region:

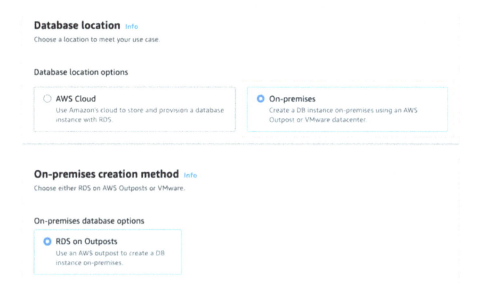

Figure 5.14 – Deploying an RDS instance to an AWS Outposts Rack deployment

Once the **On-premises** location has been selected, customers will be presented with Outposts-specific configuration options, such as a choice between subnets within the AWS Outposts Rack deployment or using **Customer-owned IP (CoIP)** addresses. This can be seen in the following figure:

Figure 5.15 – Selecting an address type for a new RDS instance

Amazon Elastic Map Reduce (Amazon EMR)

Amazon EMR is the industry-leading cloud big data solution for petabyte-scale data processing, interactive analytics, and ML that uses open source frameworks such as `Apache Spark`, `Apache Hive`, and `Presto`. Amazon EMR clusters can be deployed within Outposts for big-data processing.

Note that if service link connectivity is lost, an EMR cluster will continue running in a degraded fashion. While disconnected, customers will not be capable of creating new clusters, adding steps, or checking step execution. Furthermore, the delivery of CloudWatch events and metrics will be delayed. Should a disconnect event persist for more than a few hours, any clusters that do not have terminate protections enabled could be terminated.

AWS App Mesh

AWS App Mesh is a service mesh that's designed to ease service monitoring and controls. Service-to-service communications are routed through the service mesh. This enhances flexibility and functionality and maximizes control over application communications. AWS App Mesh provides low-latency service mesh connectivity for data and applications that exist on-premises and within the Outposts deployments. Should network connectivity to the AWS region be interrupted, App Mesh Envoy proxies will continue to run but no changes can be made until the connectivity is restored.

Amazon Elastic Container Service (Amazon ECS)

ECS is a fully managed container orchestration service that simplifies the deployment, management, and scaling of containerized applications. Customers can launch non-Fargate versions of Amazon ECS within an AWS Outposts Rack deployment for full-scale container orchestration:

Cluster name

op-ecs-cluster-jaynaves

There can be a maximum of 255 characters. The valid characters are letters (uppercase and lowercase), numbers, hyphens, and underscores.

▼ **Networking** Info

By default tasks and services run in the default subnets for your default VPC. To use the non-default VPC, specify the VPC and subnets.

VPC

Use a VPC with public and private subnets. By default, VPCs are created for your AWS account. To create a new VPC, go to the VPC Console ☑.

vpc-04
jaynaves-create-vpc-cidr-ec2-VPC   ▼

Subnets

Select the subnets where your tasks run. We recommend that you use three subnets for production.

Choose subnets   ▼

subnet-018   public ✕
jaynaves-create-vpc-cidr-ec2-OutpostsSubnet
us-west-2a   10.50.1.0/24

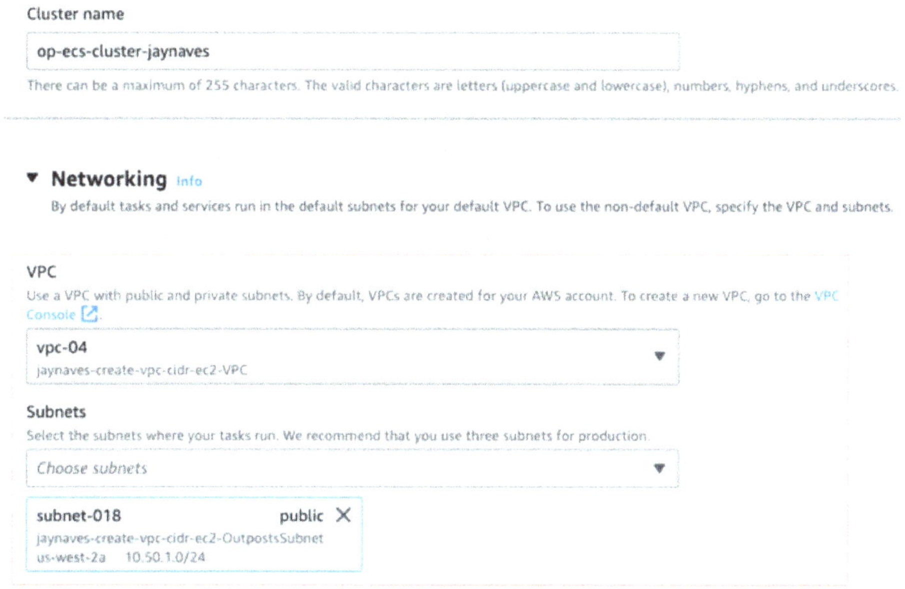

Figure 5.16 – Launching an ECS cluster within a deployment by selecting an Outposts subnet

The supplemental services of Amazon **Elastic Container Registry** (**ECR**), AWS **Identity and Access Management** (**IAM**), **Network Load Balancer** (**NLB**), and Amazon Route 53 require AWS Region connectivity. The lack of access to these supplemental services means that no new clusters can be created, no new actions can be performed on existing clusters, instance failures will not be automatically replaced, and CloudWatch logs and event data will not propagate.

Amazon Elastic Kubernetes Service (Amazon EKS)

EKS is a managed Kubernetes service that runs in AWS regions as well as on-premises. The Kubernetes control plane can be deployed into the AWS region or on-premises within AWS Outposts but both deployments are fully managed by AWS. The control plane manages key components of an EKS cluster such as application availability, cluster data, and scheduling for containers.

Amazon EKS on Outposts enables customers to leverage on-premises cluster capacity in a low-latency manner using the same tools and APIs that customers use to access Amazon EKS in the AWS cloud. Customers can deploy Amazon EKS cluster nodes onto both virtual machines and bare-metal servers within their AWS Outposts Rack deployments.

Customers deploy their Amazon EKS worker nodes into an AWS Outposts Rack deployment by using `self-managed node groups` versus `managed node groups`. Compared to running EKS in-region, this shifts more of the responsibility for updating nodes to the customer whenever new Amazon EKS-optimized AMIs are released for use.

Customers can create Amazon EKS clusters in two ways:

- **Extended clusters**: The Kubernetes control plane runs in an AWS region but the nodes run on AWS Outposts. This approach is generally chosen to minimize the compute capacity footprint required for an AWS Outposts Rack deployment. The primary consideration is the stability of the service link connection as any disconnection of a cluster's nodes from its control plane is likely to disrupt cluster manageability and application availability.

- **Local clusters**: The Kubernetes control plane and nodes run on AWS Outposts. This is typically done to minimize the impact of service link disconnections:

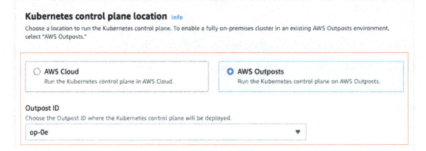

Figure 5.17 – Selecting an EKS control plane location

High availability (HA)

Multiple approaches are possible to ensure the appropriate level of availability for a customer's use case.

HA within a deployment

All AWS Outposts Rack deployments offer local HA through the use of redundant networking and power equipment. This redundant Outposts equipment can be paired with dual customer-provided power sources and redundant network connectivity for even greater resilience. AWS recommends combining the built-in network redundancy with customers' network redundancy to further ensure constant connectivity for the Outpost.

Additional built-in and always active capacity can be provisioned within a deployment, offering further improvements to availability. AWS offers production-ready deployment configurations that support an N+1 capacity level for EC2 instances.

AWS Outposts Rack deployments provide two instance placement options for minimizing the impact of hardware failures:

- **Dedicated hosts** – This offers customers the ability to use auto-placement configurations to dictate that instances be deployed onto a specific host or any available host matching certain requirements. This is done using host affinity.
- **Placement Groups**: These work similarly to the same feature available within a region, with some modifications.

Cluster placement groups ensure instances wind up on the same physical server. This is normally done to ensure the lowest possible latency between two instances.

Customers with more than one rack in their deployment can use a spread placement group at the rack level to ensure that the instances in a group are spread across multiple racks.

The spread placement strategy of partition is used to distribute multiple instances across multiple racks within partitions. Single partitions can house multiple instances and customers can target specific partitions for instance placement or they can use automatic distribution to spread the instances across the partitions. The following figure shows an example of this approach:

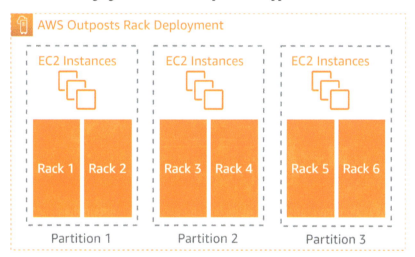

Figure 5.18 – A spread placement group using targeted partitions

Additionally, an AWS Outposts Rack deployment offers a unique spread placement strategy at the host level. This is meant for single-rack deployments so that customers can safely spread deployments of instances across physical servers within a single rack.

Customers should utilize Amazon CloudWatch to monitor capacity availability metrics and set alarms for application health, create CloudWatch actions to perform automated recovery, and monitor the utilization of your Outpost deployment. When instances are marked as unhealthy, they need to be migrated to a healthy host. These CloudWatch capabilities allow customers to enable automated migrations of unhealthy hosts based on instance status checks. Auto-scaling groups can also be enabled to achieve a similar effect.

HA across deployments

Statistically, it is quite rare for hardware failure to be the source of an outage that impacts application availability. Most are caused by misunderstandings of logical dependencies during maintenance. A good way to prevent this impact is to have a completely separate Outposts Rack deployment that is tied to a different parent region. These twin deployments can, and often do, reside physically within the same site – although the diversity of utility providers such as power and internet connectivity should be maintained between them:

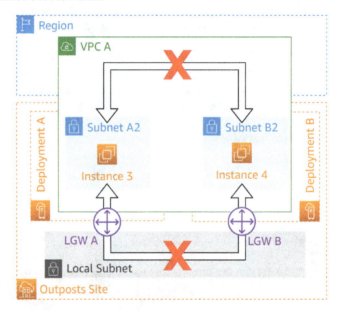

Figure 5.19 – Communication path restrictions for two deployments in the same VPC

If customers choose to take this approach, they must take the following restrictions into account:

- Network traffic from resources on two Outposts cannot traverse the AWS region (see *Figure 5.19*)
- Network traffic from resources on two Outposts cannot traverse the customer network unless the resources are on different VPCs (see *Figure 5.20*):

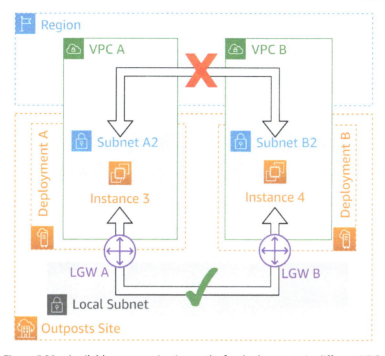

Figure 5.20 – Available communication paths for deployments in different VPCs

In either case, all Outpost-to-Outpost traffic is blocked if it traverses the region. This block is in place to prevent redundant network egress charges from both the originating AWS Outposts Rack deployment and from the region.

Local VPC CIDR routes will always direct traffic between resources in the same VPC through the region and therefore will not be allowed because of this block. Due to this limitation, traffic between two AWS Outposts Rack deployments is only allowed to traverse the customer's local network via the LGWs, and even then, only when separate VPCs are in use.

Security

AWS Outposts builds upon the `AWS Nitro System` to provide customers with enhanced security mechanisms that protect, monitor, and verify your Outpost's instance hardware and firmware. This technology allows resources needed for virtualization to be offloaded to dedicated software and hardware, which provides an additional layer of security through the reduction of the attack surface. The Nitro System's security model prohibits admin access to eliminate the possibility of human error or bad-actor tampering.

Similarly, the traditional `AWS Shared Responsibility Model` is also built upon by AWS Outposts but updated to fit the service's on-premises aspects. In the following figure, take note of how the dividing line for a customer's responsibilities shifts to cover additional responsibilities. Examples include elements such as physical security and access control, environmental concerns, network redundancy, and capacity management:

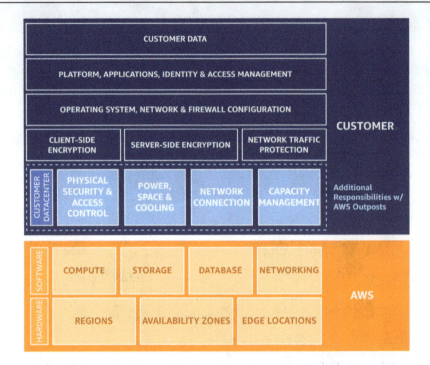

Figure 5.21 – Shared security model modified for AWS Outposts

The EBS volumes and snapshots that are used within an AWS Outposts Rack deployment are encrypted by default using AWS **Key Management Service (KMS)** keys. Similarly, all data transmitted between a customer's deployment and the region is encrypted. Customers are responsible for encrypting data in transit between the LGW and its attached local network using protocols such as **Transport Layer Security (TLS)**.

A primary reason that disconnecting the service link from the region is problematic is the way such communications are encrypted. All communications to an AWS Outposts deployment must be signed with an IAM principal's access key and a secret access key or be temporarily signed using the **AWS Security Token Service (AWS STS)**. Furthermore, the publication of VPC flow logs to CloudWatch, S3, or Guard Duty will cease to function properly during disconnected events.

A local security feature the Nitro hypervisor provides is the automatic scrubbing of EC2 instance resources after use. This includes all memory allocated to an EC2 instance by setting each bit to zero whenever it is stopped. When it is terminated, a similar wipe of its attached block storage also occurs.

Physical security

Delivery and installation are performed through secure channels. Furthermore, any replacement parts are delivered through these same secure channels. Note that replacements are always provided in the event of failed hardware – no network switches or servers will be repaired after installation, only replaced.

Using AWS Outposts Server

AWS Outposts servers are designed in smaller 1U and 2U form factors that fit inside standard 19" full-width racks and provide access to the same AWS services, infrastructure, tools, and APIs as AWS Outposts racks. While AWS takes ownership of the Outposts rack installation, for Outpost servers, they provide customers with an easy-to-use installation interface so that they can handle the physical installation of the Outposts server themselves.

Ordering

Once a customer receives an Outposts server, they will be responsible for racking it, providing it power and network connectivity, powering it on, and connecting to the server via an included USB cable for initialization and authorization.

The server options have fewer power draw requirements than Outposts racks but do come with both DC and AC connectivity options. The network throughput requirements are also less than AWS Outposts Rack deployments but both server options come with 10 Gbps networking capabilities. The EC2 instances are still AWS Nitro backed and contain the Nitro Security Key.

The 1U option weighs 26 pounds and is 1.75" tall, 17.5" wide, and 30" deep and provides C6gd compute-optimized EC2 instances running Arm-based AWS silicon called Graviton2 with up to 64 vCPUs, 128 GiB of RAM, and 2x 1,900 GiB NVMe SSD. The 2U option doubles the vCPUs, RAM, and NVMe at 36 pounds and 30" deep. It provides C6id compute-optimized EC2 instances using the Intel Xeon Ice Lake processor.

Customers can choose from different configurations available for Outposts servers, such as storage and EC2 configurations. Customers are required to install and maintain the installation of the Outposts servers so that AWS does not charge the customer for that. There are all upfront, partial upfront, and no upfront options for a 3-year commitment term. Monthly charges will occur for the partial and no upfront payment options. The agreed-upon upfront charges get counted 24 hours after the Outposts server is provisioned for use.

Customers looking to leverage Outposts' capabilities in a scenario that requires less in the way of compute and storage capacity or in a scenario that does not allow for a full-size rack install can consider using one of the two Outposts server form factors. Customers that have a workload need for limited AWS resources and have space available in an existing server rack make good use cases for Outposts servers. Also, the fact that Outposts servers can be combined for additional throughput back to the AWS region is a feature that allows customers to expand their Outposts server footprint to add additional throughput, resource capacity, and resilience.

Physical networking

Outposts servers require two physical connections to perform properly and the cables needed for this connectivity are provided to customers with their Outposts server. The two connections are used to segment traffic according to traffic class and are not intended for redundancy.

These connections are not required to be on a common network. The cable provided for this use is a QSFP+ to 4xSFP+ breakout cable. The QSFP+ interface plugs into port 3 of the Outposts server. The SFP+ interfaces plug into the customer switch side with SFP+ interface 1 used for traffic labeled 1: LNI traffic and SFP+ interface 2 for traffic labeled 2: service link traffic. Note that SFP+ interfaces 3 and 4 are not currently used in an Outposts server deployment.

The customer switch port for the service link should be an untagged access port with routing capabilities out to service link endpoints in the AWS region and the Outposts registration endpoint in the AWS region. This connection must also have public DNS available for resolution of service link endpoints and for proper registration with the Outposts registration endpoint. Note that TCP and UDP ports 443, along with UDP port 53, need to be allowed and that NAT is allowed between the Outposts server and the registration endpoint.

Service link

Outposts Server provides a local Layer 2 connection compared to requiring **Border Gateway Protocol (BGP)** through a routed network for the service link.

The maximum throughput over any one Outposts Server service link is 500 Mbps. Customers can leverage multiple Outposts servers to increase the effective throughput back to the region according. For example, three Outposts servers could be used to achieve 1.5 Gbps throughput to the region in aggregate.

The customer switch port for the LNI traffic should be a standard access port to a VLAN on the customer network. Customers with more than one VLAN must configure the port to accept multiple MAC addresses as each EC2 instance launched will use a unique MAC address, and also configure all ports as trunk ports. Note that you will need to use trunk mode for your LNI; you will be responsible for any VLAN tagging needed at the EC2 operating system level as Outposts servers do not perform VLAN tagging.

Logical network interfaces (LNIs)

Unlike the shared LGW found in an AWS Outposts Rack deployment, Outposts Server enables each EC2 instance to communicate directly with on-premises infrastructure by way of an LNI:

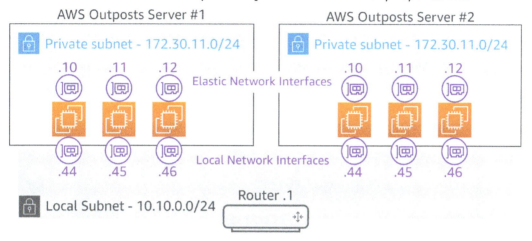

Figure 5.22 – LNIs on an AWS Outposts server

The LNI having direct communications through the LAN means customers don't need to set up routing or gateways for the Outposts servers to communicate locally. Customers are responsible for ensuring that proper multi-network routes exist on the EC2 instances to allow for communications through the LNI and also through the standard **Elastic Network Interface** (**ENI**), which is automatically provided to an EC2 instance for communication to the VPC. The ENIs on an Outposts server behave identically to the way ENIs work in-region. The ENIs provide connectivity to public or private VPC endpoints in the parent region.

High Availability (HA)

AWS Outposts servers offer HA through the use of redundant power equipment that can be paired with dual customer-provided power sources for even greater resilience. Power is critical to any compute resource but network connectivity is also critical for Outposts deployments as management, monitoring, and service operations need connectivity to the anchor AZ. However, Outposts servers do not contain redundant physical networking ports, so AWS recommends that customers ensure maximum network redundancy as far as they can to minimize network disruptions to the Outpost.

Customers can improve upon the availability of their Outposts servers' workloads by deploying additional Outposts servers. AWS recommends that customers deploy N+1 instances for each instance family that's used when there is sufficient additional capacity. In the case of Outposts server hardware issues, this N+1 strategy will enable customers to use CloudWatch actions to deploy recovery and failover mechanisms. Also, customers can alert on metrics such as capacity issues and application health with CloudWatch alarms.

Customers deploying multiple Outposts should choose different AZs for their additional deployments and provision applications onto multiple Outposts to further increase application availability in the case of AZ failures.

Service availability

Amazon EC2, Amazon Virtual Private Cloud, Amazon **Elastic Container Service (ECS)**, and AWS IoT Greengrass are the only services that are supported on AWS Outposts Server at present.

Amazon Elastic Compute Cloud EC2

The selection of instance types is limited to the following on AWS Outposts servers:

- 1U option

  - C6gd.16xlarge w/AWS Graviton2 CPU (ARM)

- 2U option

  - C6id.16xlarge

  - C6id.32xlarge w/Intel Xeon Ice Lake processor

Amazon Elastic Container Service (Amazon ECS)

Customers can launch non-Fargate versions of Amazon ECS on AWS Outposts servers for full-scale container orchestration.

Note that the supplemental services of Amazon **Elastic Container Registry (ECR)**, AWS **Identity and Access Management (IAM)**, **Network Load Balancer (NLB)**, and Amazon Route 53 require AWS region connectivity. The lack of access to these supplemental services means that no new clusters can be created, no new actions can be performed on existing clusters, instance failures will not be automatically replaced, and CloudWatch logs and event data will not propagate.

Storage

Amazon EBS volumes are not supported on Outposts servers; only instance store volumes are available. Similar to the in-region instance store, the instance store data persists after a reboot but not an instance termination. This means customers must use either stateless configurations or must back up instance data to a persistent destination such as Amazon S3 or on-premises storage.

Note that there is no local AMI caching mechanism, so the EC2 AMI will be brought in from the region on every EC2 instance launch. This needs to be taken into consideration when you're deciding on the networking needs for an on-premises deployment of an Outpost server.

Summary

In this chapter, we delved into the intricacies of two managed hybrid cloud offerings, AWS Outposts Rack and AWS Outposts Server, each of which offers distinct mechanisms for integrating cloud services with local infrastructure. We sought to provide a holistic understanding of how organizations can seamlessly bridge their on-premises environments with the power of AWS services, ensuring a cohesive hybrid cloud experience.

The section devoted to AWS Outposts Rack unveiled the critical considerations in both physical and logical realms. A comprehensive overview of the AWS services available on AWS Outposts Rack showcased how customers can leverage extended cloud capabilities closer to home. We also explored the importance of high availability and security measures, ensuring resilience and protection for mission-critical workloads. The journey continued with AWS Outposts Server, where we dissected its physical and logical attributes and how they differ from AWS Outposts Rack.

In the next chapter, we will investigate how to bring the capabilities of an AWS region closer to the edge using AWS Local Zones.

6

Lowering First-Hop Latency with AWS Local Zones

In the world of cloud computing, latency is the invisible force that can determine the success or failure of an application. From augmented reality experiences requiring instantaneous feedback to mission-critical industrial applications where every millisecond counts, reducing latency is pivotal. As we journeyed through AWS's vast edge computing landscape, it became evident that proximity to the end user is a game-changer. Enter AWS Local Zones.

Think of Local Zones as extensions of AWS's massive infrastructure regions, tactically positioned closer to specific urban centers and densely populated areas. While the primary AWS Regions might be well-suited for hosting applications with a global user base, they aren't always optimal for the ultra-low latency requirements of a specific locale.

This chapter will explain AWS Local Zones and review architectural patterns that have proven successful with customers through the following sections:

- Introduction to AWS Local Zones
- Connecting on-premise networks to AWS Local Zones with AWS Direct Connect
- Routing internet traffic into AWS Local Zones

Introduction to AWS Local Zones

AWS Local Zones are extensions of an AWS Region into a metro area that does not have a full-scale region of its own. They are tied to a specific availability zone within their parent region through AWS's robust, high-bandwidth private network.

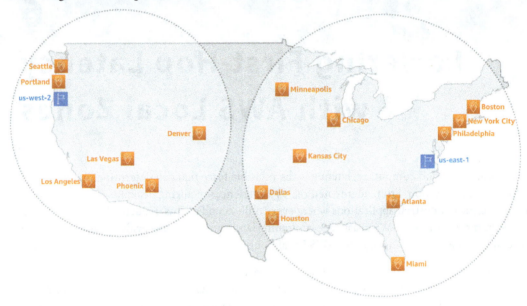

Figure 6.1 – AWS Local Zones are each tied to a parent region

This means application elements running in AWS Local Zones benefit from quick, safe, and integrated access to a full suite of services in the main region.

AWS Local Zones are available in 33 metropolitan areas around the world—17 outside of the US (Auckland, Bangkok, Buenos Aires, Copenhagen, Delhi, Hamburg, Helsinki, Kolkata, Lagos, Lima, Manila, Muscat, Perth, Querétaro, Santiago, Taipei, and Warsaw) and 16 in the US. AWS has plans to expand into 18 new locations in 16 countries, including Australia, Austria, Belgium, Brazil, Canada, Colombia, Czech Republic, Germany, Greece, India, Kenya, Netherlands, Norway, Portugal, South Africa, and Vietnam.

The AWS Local Zones User Guide maintains an up-to-date list.

Customer needs

While there are many use cases for AWS Local Zones, the drivers behind all of them boil down to three things.

Reduced latency

In today's digital landscape, many tasks, including medical image processing, real-time gaming, telco virtualization, and enterprise transactions, necessitate fast access to compute power with minimal lag.

Figure 6.2 – Customers in Texas face latency challenges to any full region

For users in the same metropolitan vicinity, AWS Local Zones provide low, single-digit millisecond latency. The latency between Local Zones and AWS Regions or Local Zones and on-premises environments varies, though it remains consistently under 10 milliseconds. The precise latency value, however, can vary based on factors such as the distance to the Local Zone and the chosen connection method, whether public internet, VPN, or AWS Direct Connect.

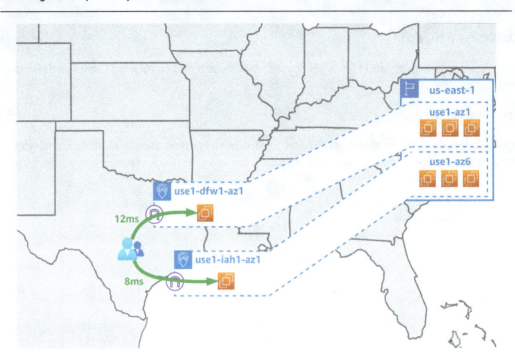

Figure 6.3 – Customers in Austin can benefit from the zones in Dallas and Houston

Many customers have transitioned their on-site workstations to AWS, reaping benefits in terms of content development speed, enhanced security, and efficient operations. Some of these workstations replace high-powered desktops with sizable GPU cards for tasks such as media content development. Especially in these cases, an optimal remote working experience demands less than five milliseconds of latency to their AWS-based virtual instances. With AWS Local Zones, these latency requirements can be accommodated. For instance, Netflix leverages G4dn-family EC2 instances running in AWS Local Zones for their media editing and other content creation tasks.

Data localization

Certain industries, such as healthcare, finance, energy, or the public sector, often need to store data within specific geographic confines for regulatory reasons. Local Zones bring AWS closer or within a customer's geographic boundary in a fully AWS-owned and operated mode and can therefore help them meet data residency requirements.

The **General Data Protection Regulation (GDPR)** is a stringent set of rules and penalties instituted by the European Union to safeguard the personal data and privacy of EU citizens. Its inception has steered organizations worldwide across all industries to re-assess their data management practices. The next figure illustrates which EU member countries have access to full AWS regions or AWS Local Zones. It also highlights those countries where AWS Outposts is a necessity:

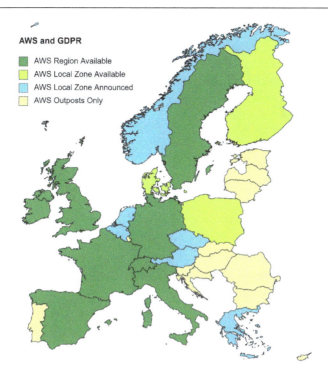

Figure 6.4 – AWS Regions, AWS Local Zones, and AWS Outposts by EU member country

Keep in mind that the applicability of any service for a given compliance regime depends on many factors. It is important to work closely with your compliance and information security teams when choosing the AWS Local Zone location in which to deploy your regulated workloads.

Some customers, such as those in the financial services industry, benefit from AWS Local Zones in multiple ways. Payment processing requires consistent low-latency data transfer. Otherwise, the user experience suffers, which can translate to direct and measurable financial loss. Simultaneously, there are often regulatory requirements imposing data to be handled and/or stored in a specific geographic location.

Such dual requirements are even more stringent in regulated gaming. For example, the United States Federal Wire Act makes it illegal to facilitate bets or wagers on sporting events across state lines. This regulation requires that operators make sure that users who place bets in a specific state are also within the borders of that state. At the same time, the timing of transactions is even more important—gamblers and casino owners alike aren't known for their patience.

Streamlined migrations

Migrating to the cloud and updating your infrastructure can be tricky. The simultaneous migration of interconnected applications can be challenging; thus, many choose not to entirely abandon their infrastructures in one shot. An incremental approach to lift-and-shift migrations was typical even before AWS Local Zones were introduced.

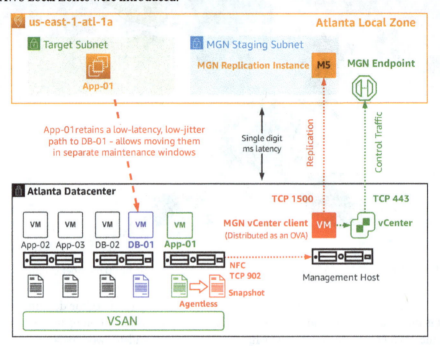

Figure 6.5 – Migrating from VMware to EC2 in a Local Zone with AWS Application Migration Service

However, challenges emerge when customers have complex, mission-critical enterprise applications. Tight latency requirements between dependencies are common, yet the closest region is sometimes 40 ms away. In such cases, the entire thing must be shut down and cut over in one motion.

AWS Local Zones and Direct Connect combined facilitate low-latency/low-jitter connectivity between cloud-based resources and systems still located on-premises. Establishing a hybrid environment that includes AWS Local Zones in its architecture gives customers a middle-ground option between on-premises and a full AWS Region.

AWS Outposts versus AWS Local Zones

When diving into AWS's infrastructure offerings, it is crucial to understand the operational overhead associated with each service. AWS Outposts and AWS Local Zones, while catering to similar requirements, differ considerably in their operational demands. When does it make sense to use one over the other?

Hardware management

AWS Outposts involves direct interaction with physical hardware. Customers are responsible for providing suitable space within their data centers, ensuring there's adequate power, cooling, and network connectivity. While AWS will handle maintenance and repairs, on-site access for AWS personnel or specific instructions for the customer might be needed, especially during hardware failures.

Being an extension of AWS's infrastructure, Local Zones do not demand customers to handle any physical hardware. The responsibility for maintaining the infrastructure remains entirely with AWS.

Network configuration

Ensuring seamless connectivity between the Outposts rack and the on-premises network is the customer's responsibility. They need to configure local network settings, integrate with existing network setups, and ensure secure and high-bandwidth connections to AWS Regions.

Since Local Zones are designed to be extensions of AWS Regions, AWS manages the majority of the networking aspects. Customers focus on their VPC settings, just as they would in a typical AWS environment.

Capacity management

Outposts' capacity is finite. As it is a physical installation, any expansion requires careful planning and potentially additional hardware. Customers need to monitor usage and predict future needs.

The elasticity inherent to the cloud is present in Local Zones. While there might be soft limits, scaling resources is more straightforward, not requiring hardware interventions.

Security

Since Outposts hardware resides within the customer's facility, ensuring that physical access controls, surveillance, and auditability become their responsibility. Furthermore, customers must ensure robust firewall configurations for an AWS Outposts service link, as well as safeguard against on-premises network vulnerabilities.

AWS Local Zones, on the other hand, maintain a cloud-centric security paradigm, primarily focusing on data and access management. The shared responsibility model customers are used to in-region remains unmodified.

AWS Dedicated Local Zones

AWS Dedicated Local Zones is a new offering built at the request of two types of customers. First, it helps those with particularly stringent constraints imposed by compliance regimes that require physically separate infrastructure. Second, it is an option for customers who like the idea of a fully managed service that includes the physical layer but do not currently have an AWS Local Zone in their area.

The infrastructure involved can be deployed to any data center in which AWS is able to ensure they can deliver on their commitments around physical security, connectivity, and so forth. Because each implementation is by definition a customized offering, you `must contact AWS` to establish whether this service can meet your needs.

AWS Local Zones pricing

When considering pricing for AWS Local Zones, it's important to note that costs for instances and other AWS services within a Local Zone may vary from their corresponding prices in the main region. For EC2 instances in Local Zones, options such as On-Demand, Spot Instances, and Savings Plans are available.

Data transfer fees within AWS Local Zones align with the rates of availability zones in the primary region. For example, transferring data between EC2 instances in the Los Angeles zone and Amazon S3 in the main region, US West (Oregon), incurs no charges. However, moving data to and from an EC2 instance in the Local Zone and an EC2 in the main region is priced at $0.01/GB each way.

For more specific, up-to-date information regarding pricing in AWS Local Zones, please refer to the `AWS Local Zones pricing` page on the AWS website.

Latency to AWS Local Zones

Tools such as the AWS latency tester [`https://aws-latency-test.com`] help predict the latency benefits that a given device will experience when connecting through an AWS Local Zone vs a region:

⊞ **Local Zone Latency median (ms)**

Local Zone	ap-northeast-1-tpe-1a	ap-south-1-ccu-1a	ap-south-1-del-1a	ap-southeast-2-per-1a	eu-central-1-ham-1a	eu-central-1-waw-1a
Location	Taipei, Taiwan	Kolkata, India	Delhi, India	Perth, Australia	Hamburg, Germany	Warsaw, Poland
Parent region	Asia Pacific (Tokyo)	Asia Pacific (Mumbai)	Asia Pacific (Mumbai)	Asia Pacific (Sydney)	Europe (Frankfurt)	Europe (Frankfurt)
Round trip (msec)	132	46	29	320	170	170

Figure 6.6 – AWS Local Zone latency test output

This figure shows a small snippet of the output of this tool. It will geolocate the client's IP and use a series of JavaScript functions to sample the latency to every AWS region and AWS Local Zone. Of course, the best test would be to spin up an EC2 instance or container running your application in the target zone. To do that, you will first need to opt in.

Opting into AWS Local Zones

AWS Local Zones are not enabled by default in your AWS account. You must opt into each one explicitly. This can be done in the AWS Management Console by navigating to **VPC > Dashboard** and clicking on the **Zones** link, as shown in the following figure:

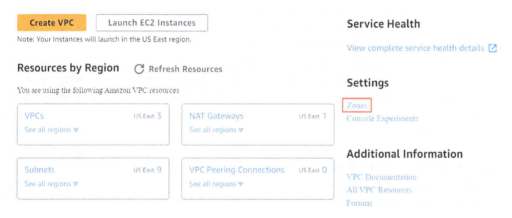

Figure 6.7 – The VPC dashboard in the AWS Management Console

This must be done on a per-region basis and can become quite tedious, especially once you have multiple AWS accounts involved in an organization. It is best to automate this process, and there are multiple ways to go about this. With the AWS CLI, execute the following command to see which AWS Local Zones are children of a given region in an easy-to-read, tabular format:

```
aws ec2 describe-availability-zones \
--region us-west-2 \
--filters Name=zone-type,Values=local-zone \
--all-availability-zones \
--query AvailabilityZones[*].[*] \
--output text | sort
```

The next figure shows part of the output from the preceding command. The columns shown here are, in order, Group Name, Zone Type, Parent ZoneName, and Parent ZoneId:

```
us-west-2-den-1 us-west-2-den-1 local-zone     us-west-2d     usw2-az4
us-west-2-las-1 us-west-2-las-1 local-zone     us-west-2c     usw2-az3
us-west-2-lax-1 us-west-2-lax-1 local-zone     us-west-2b     usw2-az2
us-west-2-lax-1 us-west-2-lax-1 local-zone     us-west-2d     usw2-az4
us-west-2-pdx-1 us-west-2-pdx-1 local-zone     us-west-2c     usw2-az3
us-west-2-phx-2 us-west-2-phx-2 local-zone     us-west-2b     usw2-az2
us-west-2-sea-1 us-west-2-sea-1 local-zone     us-west-2a     usw2-az1
```

Figure 6.8 – Child AWS Local Zones of us-west-2 (Oregon)

> **Availability zone identifiers: ZoneId vs ZoneName**
>
> All availability zones, including Local Zones, have two identifiers—a ZoneId and a ZoneName. The ZoneId is unique and consistent at all times. However, the ZoneName is a dynamic alias pointing to the static ZoneId. These aliases are randomly generated for an account whenever it is created.
>
> Therefore, what you call us-west-2a could be what another customer calls us-west-2b. This is done to balance the utilization of the zones. Without this, the a zone in each region would be heavily overutilized compared to the others.

Each AWS Local Zone is tied to not only a parent region but also to a specific availability zone. It is, therefore, important to keep track of the physical ZoneId as we proceed.

The following command will opt into a given AWS Local Zone for the account in question. We need to do this using the Group Name, which is an identifier for all AWS Local Zones in a given metro area. Some, such as the one in Los Angeles, have multiple zones, similar to a region:

```
aws ec2 modify-availability-zone-group \
--region us-west-2 \
--group-name us-west-2-lax-1 \
--opt-in-status opted-in
```

Note that once you opt into an AWS Local Zone, you must contact AWS support to opt out.

Connecting on-premises networks to AWS Local Zones

For use cases where resources in an AWS Local Zone need to communicate with resources in an on-premises data center in that same metro area, there are two basic approaches. The first is a physical connection via AWS Direct Connect and the second is a virtual connection via a VPN over the public internet. Both approaches have benefits and drawbacks, which must be weighed in light of the requirements and constraints that a given project is operating within.

AWS Direct Connect

AWS Direct Connect is a service that provides dedicated network connections from on-premises environments (such as a corporate data center) to AWS. Instead of using the public internet for AWS data transfer, Direct Connect offers a private, high-bandwidth, and consistent network experience. It often results in reduced network costs and increased bandwidth throughput and provides a more consistent network experience than typical internet-based connections.

An `AWS Direct Connect Location` refers to the specific physical place where AWS has partnered with colocation facilities around the world to provide access points to its network. These locations act as network access points into AWS, allowing users to establish dedicated network connections from their on-premises environments to AWS.

Note that while any given location is tied to a specific AWS Region, this is a management construct only. All AWS Direct Connect Locations connect to all AWS Regions, including AWS GovCloud Regions. They also connect to all AWS Local Zones as well as to other AWS Direct Connect Locations.

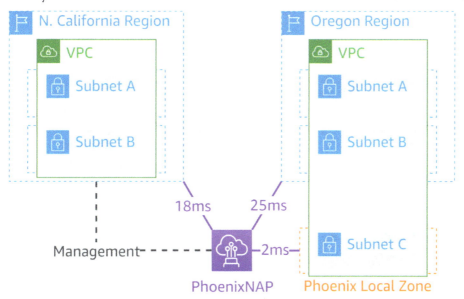

Figure 6.9 – Latency from the AWS Direct Connect Location PhoenixNAP

This means traffic from your on-premises data center will be routed directly to the AWS Local Zone in the same metro area, even if the AWS Direct Connect Location you are connected to is logically tied to a different parent region. The previous figure illustrates the logical situation.

Physical networking—dedicated connection

An AWS Direct Connect Location physically consists of an AWS-managed footprint within a cage of a major colocation facility. Due to this, it is not uncommon that customers themselves have a cage in the same colocation facility. In these cases, a dedicated connection may be ordered. The following figure shows an example of this configuration in the IPB facility in Berlin:

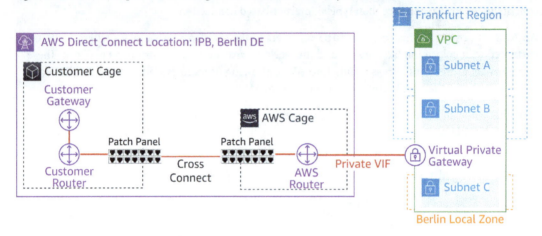

Figure 6.10 – AWS Direct Connect in Berlin

Notice the cross-connect between the customer and AWS cages. This is a physical cable that runs through a meet-me room and is ultimately terminated on patch panels on either side.

Initiating the order for a dedicated connection

The first step in the process is to navigate to the **AWS Direct Connect** page within the AWS Management Console and click **Create Connection**. Under **Connection ordering type**, select **Classic**. Next, choose the appropriate **Location** from the drop-down menu. Port speed options will vary depending upon the location chosen—only a subset will offer 100 gigabits per second.

Note that port speed for a dedicated connection is always either 1, 10, or 100 gigabits per second. For example, it is not possible to order a dedicated connection for 2.5 gigabits per second. Refer to the *Physical networking—hosted connection* section of this chapter if your needs fall into the middle ground between two speeds like this.

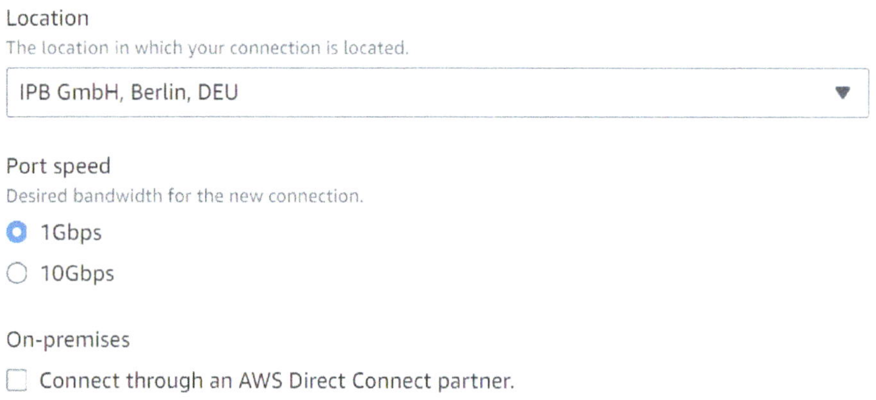

Location
The location in which your connection is located.

IPB GmbH, Berlin, DEU   ▼

Port speed
Desired bandwidth for the new connection.
🔘 1Gbps
⭕ 10Gbps

On-premises
☐ Connect through an AWS Direct Connect partner.

Figure 6.11 – Ordering an AWS Direct Connect within the same facility

In the **On-premises** section, uncheck the box that says **Connect through an AWS Direct Connect partner**. The preceding figure shows an example of this for the IPB facility in Berlin.

Under **Additional Settings**, you can choose to order a MACsec capable port or to bind this connection to an existing LAG. The latter will only be possible if a connection has already been established to your cage. The reasons why you might choose either of these options will be covered later in this chapter.

Upon submission of the request, an AWS Support case will be opened, requesting specific information such as your cage number. Once AWS Support has processed the connection request, you will receive a link for a **Letter of Authorization - Connecting Facility Assignment** (**LOA-CFA**). If you do not activate the port on your end within 90 days, billing will automatically start 90 days after the LOA-CFA is issued unless you cancel the process during that time.

Running the cross connect

Once you have an LOA-CFA, you need to have the facility run the cross connect. Procedures for accomplishing this vary by facility, but they typically involve opening a ticket with the facility in question to have their staff connect the appropriate cable type. They will also test the physical path end-to-end—something particularly important with fiber optic cable runs. Check the `Cross connects` section of the `AWS Direct Connect User Guide` for the most up-to-date information regarding a particular facility.

Note that it is best to involve the facility early on, long before the LOA-CFA is issued. The facility will have its own lead times to run cross-connects based on staff availability.

Once the cross connect has been run, it is the customer's responsibility to ensure their physical router has the correct transceiver type and is connected to the patch panel in their cage with appropriate cabling that matches the cross connect. Furthermore, the customer must configure their router port correctly for settings such as auto-negotiation – something that varies depending on the situation. See the `Troubleshooting layer 1 issues` section of the `AWS Direct Connect User Guide` for more details.

Physical networking—hosted connection

In many cases, the data center a customer wants to reach an AWS Local Zone from is a self-built data center in its own building or perhaps in a colocation facility that does not happen to be an AWS Direct Connect Location. In cases such as this, an AWS Direct Connect Partner can offer what is known as a hosted connection:

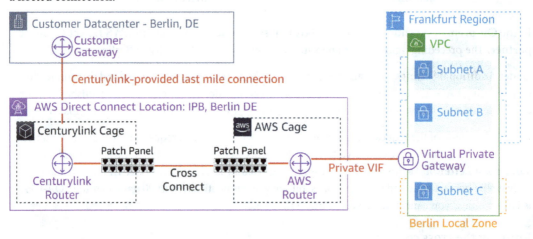

Figure 6.12 – AWS Direct Connect via partner Centurylink in Berlin

Hosted connections offer several benefits over dedicated connections:

- **Simplicity**: Customers can rely on the partner to handle the administrative and operational aspects of the physical connection, simplifying the setup process.

- **Incremental scaling**: Hosted connections offer a range of bandwidth options, starting from 50 Mbps and going up to 10 Gbps. This flexibility allows customers to choose the right capacity for their workloads and scale it up in line with throughput demand.

- **Speed of provisioning**: As they are managed by partners who have pre-established infrastructures in the appropriate locations, hosted connections tend to be set up faster.

- **Cost-effective for lower throughput**: Organizations that do not need the full capacity of a dedicated connection can opt for smaller bandwidth increments, which is typically more cost-effective even though a hosted connection costs slightly more.

Cost difference for a hosted connection

On a per-Gbps basis, hosted connections do cost slightly more than dedicated connections. However, this difference is small enough that it usually does not impact a customer's decision except at the highest of scales.

Continuing with the two examples mentioned previously of a customer connecting to the Berlin zone, let's assume the connection speed is 1 Gbps and the customer transfers 50 TB in and 50 TB out every month. The costs per month in USD are as follows:

	Port cost (1 gigabit per second)	Transfer cost (50TB)	Total
Dedicated	$219.00	$1,024.00	$1,243.00
Hosted	$240.90	$1,024.00	$1,264.90

Figure 6.13 – Simple comparison of hosted and Direct Connection costs

Accounting for spikes and overhead, this much transfer is essentially utilizing the entire 1 Gbps connection for the entirety of the month. What happens when the customer's throughput needs scaling to 1.5 Gbps? What if the customer's utilization pattern is sporadic and they need to account for 95th percentile spikes in the 2.7 Gbps range? In those cases, the small savings from a dedicated connection are more than cancelled out by the lack of granularity.

Connection resilience

When using AWS Direct Connect to communicate with a region, AWS recommends the architecture shown in the following figure for production workloads to ensure maximum resilience:

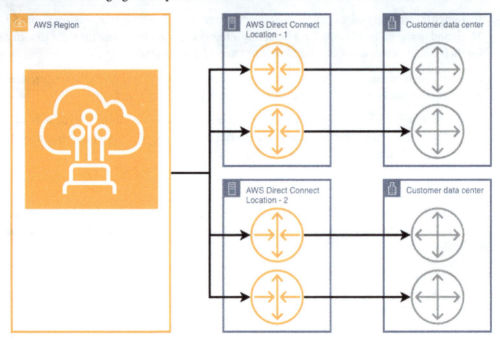

Figure 6.14 – AWS recommendation for maximum resilience with Direct Connect

At the time of writing, only the AWS Local Zone in Los Angeles has more than one availability zone in a similar fashion to a region. In situations such as this, the recommendation should be followed.

However, most AWS Local Zones are single logical entities. Further complicating this, the number of AWS Direct Connect Locations situated within a given AWS Local Zone's metro area varies. For example, Las Vegas has two AWS Direct Connect locations—SUPERNAP 8 and Databank LAS1. Helsinki, on the other hand, has only one—Equnix HE6. It would simply not be possible to implement this recommendation while retaining single-digit millisecond latency and data sovereignty across both locations—one of the locations would need to be in a different country entirely.

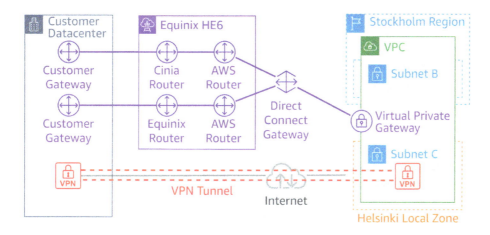

Figure 6.15 – Augmenting resilience to the AWS Local Zone in Helsinki

This figure gives examples of what could be done to maximize the resilience of connectivity in this situation. It is not uncommon for customers to run VPN server endpoints on EC2 instances such as this in AWS Local Zones, not only for resilience but also as a service they offer their customers.

Check the AWS Direct Connect Locations page on the AWS website for the latest information on availability in the area you are interested in.

Media Access Control Security (MACsec)

Media Access Control security (**MACsec**) is a protocol designed for Ethernet-based networks. Implementing MACsec in the context of AWS Direct Connect presents several benefits:

- **Data confidentiality**: MACsec encrypts packets at layer 2 (the data link layer) before they traverse the Direct Connect link. This ensures that data remains confidential while in transit, protecting sensitive information from eavesdropping or interception.

- **Integrity and authenticity**: MACsec not only encrypts the data but also authenticates it. This means that data cannot be tampered with during transit without detection. By ensuring both the integrity and authenticity of packets, users can trust that the data they receive is the exact data that was sent, untampered and from a legitimate source.

- **Protection against replay attacks**: MACsec incorporates an anti-replay mechanism, which prevents malicious actors from capturing valid data frames and re-sending them. This ensures that old or previously captured data cannot be used maliciously to impersonate valid transmissions.

- **Enhanced security posture**: By integrating MACsec with AWS Direct Connect, organizations can bolster their overall security posture. The protocol complements other security mechanisms, such as VPNs and TLS, ensuring a layered security approach. While higher-layer protocols secure data at the application or transport layers, MACsec adds another layer of protection at the data link layer.

- **Transparency**: Since MACsec operates at layer 2, it's largely transparent to higher layers in the networking stack. This means that applications and services running over Direct Connect don't need modifications to benefit from MACsec's security enhancements.

- **Low latency**: Encryption at the data link layer is efficient and introduces minimal latency. This ensures that the high-speed, low-latency benefits of Direct Connect remain largely unaffected, even when adding robust security features.

- **Compliance**: For organizations subject to stringent regulatory requirements related to data protection, incorporating MACsec can aid in compliance efforts. Ensuring data in transit is encrypted and tamper-proof can be a critical component of regulatory standards.

Routing internet traffic into AWS Local Zones

So far, we've covered use cases that involve traffic to and from an existing on-premises data center. This section will review options to address situations where the traffic originates from various locations on the internet.

Application Load Balancer

When an Elastic Load Balancer is set up in an AWS Local Zone, it is done in much the same way as it is in any standard availability zone. First, unlike in a standard region, you only need to assign one subnet. Second, only the layer 7 **Application Load Balancer** (**ALB**) variant is available. Layer 4, or **network load balancers** (**NLB**), cannot be deployed to an AWS Local Zone.

Amazon Route53 for load balancing

Although ALB addresses layer 7 load-balancing use cases, some low-latency applications that get deployed in AWS Local Zones rely on UDP-based protocols, such as QUIC, WebRTC, and SRT, which can't be load-balanced by layer 7 load balancers.

Record name	Type	Routing policy	Differentiator	Value/Route traffic to	Health check ID
wireguard.tanagra.uk	A	Weighted	10	18.130.234.187	4ec0121e-c089-40d7-b7d3-fe315b22d319
wireguard.tanagra.uk	A	Weighted	10	18.169.188.129	dd868ba3-b80a-400e-ad2d-d67a5ffe1d2b

Figure 6.16 – Amazon Route53's weighted routing policy for two WireGuard VPN servers

Fortunately, Amazon Route53 offers an alternative for such situations—weighted routing policies.

Weighted routing policies

These allow you to assign weights to multiple resources or endpoints, enabling a proportionate distribution of DNS responses among them. This is useful for load balancing, A/B testing, or gracefully migrating traffic from one resource to another. By assigning a different value to each resource, you determine the likelihood (or percentage) of traffic directed to each, giving you granular control over the distribution of incoming requests.

You can choose a value between 1 and 255 for the weight on each record. The percentage of traffic each one will respond to is evaluated on a relative basis across all of the weights. For example, the two records shown in the preceding figure each have a weight of 10. This means they are evenly weighted, so each one will respond 50% of the time. Giving each one a weight of 200 would have the same effect.

Amazon Route53's weighted routing formula

The formula used to determine how much traffic a given record receives is as follows:

$$\frac{\text{Weight of a given record}}{\text{Sum of the weights of all records}}$$

Examples are as follows:

- Record 1 – Weight 10: 3.85%
- Record 2 – Weight 250: 96.15%

Health checks

In addition to traffic distribution, Amazon Route53 health checks can be configured for each record to ensure traffic is only sent to healthy nodes. The next figure shows the health check configurations for our pair of WireGuard VPN servers in an AWS Local Zone:

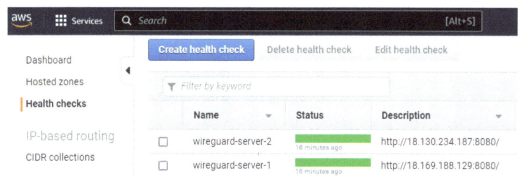

Figure 6.17 – Amazon Route53 health checks for two WireGuard VPN servers

You will notice the health check is querying `http://<ip-address>:8080/`. Because WireGuard is a UDP service on port `51820`, there is no health check type that can directly test it. Therefore, we need to configure a small web service on each WireGuard instance that answers the health check.

Let's go over instructions for setting up a basic web service on a Linux server.

On a Debian-based Linux distribution such as Ubuntu, input the following:

```
sudo apt install apache2 -y
systemctl enable apache2
systemctl start apache2
echo "<html><body>healthy</body></html>" > /var/www/html/index.html
```

On a Fedora-based Linux distribution such as Amazon Linux 2023, input the following:

```
sudo yum install httpd -y
systemctl enable httpd
systemctl enable httpd
echo "<html><body>healthy</body></html>" > /var/www/html/index.html
```

Now you need to set the Security Group for the EC2 instances to allow port `80` traffic to pass through, but you don't want to expose it to the entire internet unnecessarily. Fortunately, AWS provides managed **Prefix lists** for situations such as this:

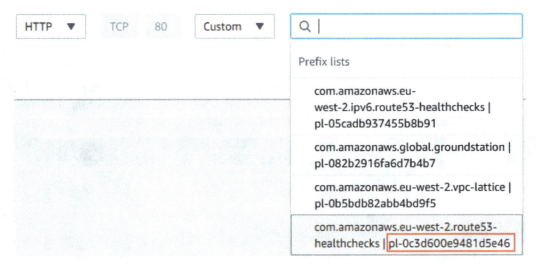

Figure 6.18 – Allowing HTTP only from the Amazon Route53 health check service

When editing the Security Group, select the **Custom** type. Then, scroll down to the bottom of the choices in the drop-down menu until you find **com.amazonaws.<region>.route-53-healthchecks**. Select it, and the prefix list ID will be entered into the rule. This will allow the health checkers in all regions to query your instance via HTTP.

By default, the health checkers are only looking for an HTTP response code 200 to consider the instance healthy. This isn't perfect, but at least proves the instance is alive on the network and the operating system is running in general.

A more sophisticated responder could be built as a Python script invoked as a service in a `systemd` `unit file`. Using `subprocess` to check the state of the VPN interface and `http.server` to reply to queries, the response could contain whatever information you want in the response, which the Amazon Route53 health check can parse to determine things such as whether the WireGuard service is functional but at capacity in terms of client connections.

DNS caching

This method of load balancing expects new connections to issue a fresh DNS query each time. Thus, it is prudent to consider the caching behavior of the client devices, as well as any DNS servers between them and Amazon Route53.

By default, the **Time To Live** (**TTL**) of an A record in Amazon Route53 is 300 seconds (5 minutes). This means clients and intermediate DNS servers should remember the last response for 5 minutes before discarding it and retrieving the record again, thus allowing the routing policy to potentially hand out a different IP this time. This value can be lowered to as little as 0, which means no caching at all should occur.

Unfortunately, not all devices obey the TTL given out by the authoritative name servers. This is particularly true of intermediate DNS servers. Some ISPs and operating systems optimize by overwriting the TTL with one they feel is more appropriate. Check with the owner of your DNS resolvers to see if this is happening or set up your own inbound private resolvers with Amazon Route53.

The tradeoff to lowering the TTL to a very low value is that you will incur more chargeable queries for Amazon Route53. As an example, let's consider an application with 1,000 devices that are online 24/7 and all query DNS as often as the TTL allows:

TTL	Queries Per Device /mo	Total Queries / mo	Cost
300	8,761	8.7 million	$4
1	2,628,288	2.6 billion	$726

Figure 6.19 – Comparison of Amazon Route53 query charges by TTL for 1,000 devices

AWS Local Zone as primary with parent region as secondary

When using an AWS Local Zone to reduce first-hop latency for users in a given metro area, using the parent region as a failover target is straightforward with AWS Route53. Separate ALBs, each with their own target groups, are created—one in the parent region and one in the AWS Local Zone:

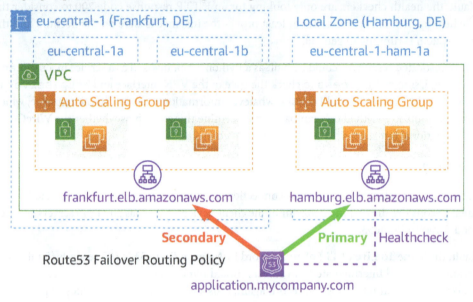

Figure 6.20 – Using Route53 to preferentially route traffic to the Hamburg zone

First, within a Route53 Hosted Zone for your public-facing DNS domain, create a health check that points to the preferred ALB.

Second, create two ALIAS records, one pointing to each ALB. Both records should have the same record name (i.e. myapp.mycompany.com). Use a failover routing policy with the record pointing to the preferred ALB set as primary and the other record set as secondary. When you create the primary record, attach the health check created earlier. See the following figure for an example of this configuration:

Record name	▽	Routing policy	Differentiator	Value/Route traffic to	Health check ID
application.tanagra.uk		Failover	Primary	dualstack.lax-1a-alb-2065223023.us-west-2.elb.amazonav	a86d81bb-0342-4ff7-9785-7e11d5b4eee4
application.tanagra.uk		Failover	Secondary	dualstack.us-west-2-alb-857795129.us-west-2.elb.amazon	-

Figure 6.21 – Route53 Primary/Secondary failover example

Using AWS Global Accelerator

Where AWS Route53 manipulates DNS responses, AWS Global Accelerator utilizes IP anycast to improve the availability and performance of applications. AWS Global Accelerator uses static anycast IP addresses to route user traffic to the optimal AWS endpoint based on proximity, health, and routing policies:

Figure 6.22 – AWS Global Accelerator using one IP for three locations

This means that if you have application replicas in multiple AWS Local Zones or regions, Global Accelerator will route a user's request to the closest and healthiest replica. Global Accelerator ensures that the user experiences low latency and high reliability by always being directed to the nearest and best-performing AWS location, which could be an AWS Local Zone if that is what makes sense in the situation.

Traffic between AWS Local Zones

Customers developing globally distributed architectures that leverage AWS Local Zones to extend their application's presence closer to end users sometimes wish to route traffic between two AWS Local Zones. For instance, a customer who builds a VPN service may wish to allow a customer to route in through a gateway on an EC2 instance in Miami and out a similar gateway in Boston.

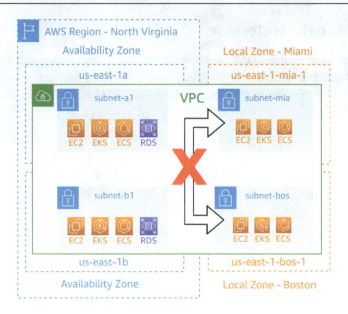

Figure 6.23 – AWS Local Zone subnets in the same VPC are blocked

In the same way two AWS Outpost rack deployments in the same VPC cannot communicate through the parent region, AWS Local Zones face the same restriction for similar reasons. The next figure illustrates the proper architecture to allow this type of communication:

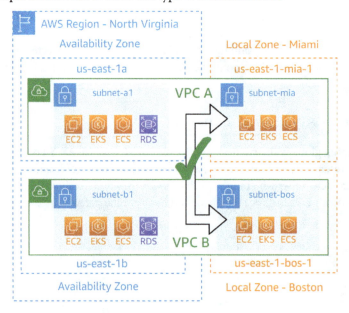

Figure 6.24 – Traffic between AWS Local Zones allowed across different VPCs

Summary

In this chapter, we covered how AWS Local Zones meet customer needs for low latency to enhance real-time data interactions, ensuring data remains within specified geographic confines and easing the migration process of on-premises applications to the cloud. It also covered its differences with AWS Outposts and how AWS Local Zones bring AWS services closer to end users without the overhead of infrastructure management.

We also covered methods for connecting on-premises data centers via AWS Direct Connect. This included a discussion of the available options depending on a customer's requirements for security, availability, performance, and cost. The chapter wrapped up by exploring how AWS Route53 and AWS Global Accelerator can be used to steer traffic from users on the internet into AWS Local Zones.

In the next chapter, we will explore a similar service that physically resides inside the 5G core of mobile carriers around the world—AWS Wavelength.

7
Using AWS Wavelength Zones on Public 5G Networks

In *Chapter 3*, we discussed how **mobile network operators** (**MNOs**) are eager to build **multi-access edge computing** (**MEC**) offerings to develop new revenue streams for their 5G infrastructure investments. However, they are not **cloud service providers** (**CSPs**). They usually do not have the in-house expertise to operationalize a customer-facing multitenant service such as this. Enter AWS Wavelength. In 2019, AWS partnered with several MNOs to build a joint service offering that extends the reach of AWS regions into their network footprint. AWS Wavelength allows customers to deploy the same compute, storage, and managed database services they are used to right at the edge of these MNOs' networks.

A Wavelength Zone is really just a special type of Local Zone. The only difference with a Wavelength Zone is that they are designed to be as close to an MNO's packet core as possible. Wavelength Zones are available in cooperation with Verizon in the US, KDDI in Japan, SK Telecom in South Korea, Vodafone in the UK and Germany, **British Telecom** (**BT**) in the UK, and Bell in Canada.

This chapter will explain AWS Wavelength Zones and review architectural patterns that have proven successful with customers. In particular, we will look at the following topics:

- Introduction to AWS Wavelength Zones
- Connecting to AWS Wavelength from mobile devices
- Extending a **Virtual Private Cloud** (**VPC**) into AWS Wavelength
- Integrating AWS Wavelength with other services

Introduction to AWS Wavelength Zones

While AWS Local Zones bring AWS services closer to end users, AWS Wavelength Zones take it one step further, embedding these services within carrier networks themselves. This tight integration ensures that 5G clients experience ultra-low latencies, while even 4G/LTE clients benefit from significantly improved response times.

AWS Wavelength Zones represent a fusion of cloud computing with mobile edge computing, setting the stage for the next generation of latency-sensitive applications.

Comparing AWS Wavelength deployments across global carriers

AWS Wavelength is a series of individual partnerships with carriers around the world.

An EC2 instance, or ECS/EKS container in an AWS Wavelength Zone, is specifically meant to service requests coming from mobile devices on that MNO's network. SLAs for a given mobile device around latency, jitter, and similar network parameters are specific to each MNO:

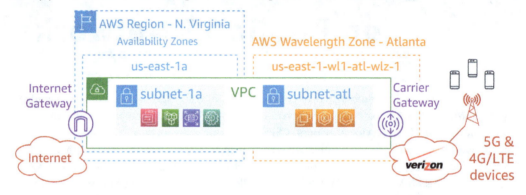

Figure 7.1 – AWS Wavelength Zone

This is because each carrier has its own unique set of customer demands and vision for how to address them.

Verizon (US)

In the US, AWS Wavelength on Verizon's expansive 5G **Ultra Wideband** (**UW**) network prioritizes bandwidth and latency, aiming to redefine the user experience. With its dense coverage in urban areas, Verizon's partnership is tailored to cater to American users, particularly those in bustling metropolitan areas. This partnership places emphasis on enhancing applications requiring single-digit millisecond latencies, such as AR/VR, game streaming, and IoT implementations.

There are AWS Wavelength Zones in 19 cities across the US. 12 of them are tied to us-east-1 (Northern Virginia) and 7 are tied to us-west-2 (Oregon).

Vodafone (Germany)

Germany, recognized for its industrial prowess, requires an edge computing solution that matches its ambitions. With AWS Wavelength's deployment on Vodafone's network, businesses in the automotive, logistics, and manufacturing sectors are in focus. The aim is to empower German industries with applications that benefit from low latency, especially given the country's emphasis on Industry 4.0 initiatives.

In Germany, Vodafone has deployed AWS Wavelength Zones in Berlin, Munich, and Dortmund.

Vodafone (UK)

In the UK, Vodafone's collaboration with AWS Wavelength supports a digital landscape that is perpetually on the move. A distinct focus is placed on sectors such as finance, healthcare, entertainment, and smart city projects. The heart of London, a global financial hub, stands to gain especially with latency-sensitive applications in real-time trading and finance.

In the UK, Vodafone has deployed AWS Wavelength Zones in London and Manchester.

KDDI (Japan)

Japan, a technological powerhouse, witnesses its digital ambitions amplified with AWS Wavelength on KDDI's network. With Japan being a leader in technological innovation, the focus shifts to sectors such as entertainment, robotics, and smart manufacturing. Applications related to streaming, especially content such as Anime and interactive gaming, are targeted to provide immersive experiences for the tech-savvy Japanese audience.

KDDI has two AWS Wavelength Zones in Japan – one in Tokyo and one in Osaka.

SK Telecom (Korea)

South Korea, a pioneer in 5G adoption, benefits from the partnership between AWS Wavelength and SK Telecom. The South Korean market, known for its high internet speeds and technological innovations, places emphasis on entertainment and smart city applications. Given the country's love for eSports and K-pop, the collaboration ensures that streaming and gaming applications have the tools to deliver unmatched quality and responsiveness.

SK Telecom has AWS Wavelength Zones in Seoul and Daejeon.

BT (UK)

BT, another major collaborator in the UK, offers a slightly different approach with its Wavelength deployment. While the emphasis remains on sectors such as finance and healthcare, BT's extensive network reach across the UK ensures that even remote areas benefit from this partnership. With BT's established enterprise customer base, Wavelength aims to simplify business transformation for legacy institutions, making digital transition smoother.

BT has one AWS Wavelength Zone in Manchester.

Bell (Canada)

Canada's vast geographical expanse requires a robust solution to tackle connectivity challenges. Bell's collaboration with AWS Wavelength aims to bring edge computing closer to Canadian users. This partnership specifically targets industries such as entertainment (given Canada's emerging importance in film and media), natural resources, and healthcare. The goal is to provide a seamless experience even in more remote parts of Canada.

Bell has one AWS Wavelength Zone in Toronto.

Connecting to AWS Wavelength from mobile devices

In previous chapters, we've focused on traffic that traverses the public internet. While an MNO has considerably more control over service quality within their own network, unfortunately, services such as AWS Global Accelerator and Amazon Route 53 geolocation don't help as much within the unique environment of a carrier's network.

Enabling AWS Wavelength Zones

As with AWS Local Zones, by default all AWS Wavelength Zones are disabled. They can be enabled in the same way:

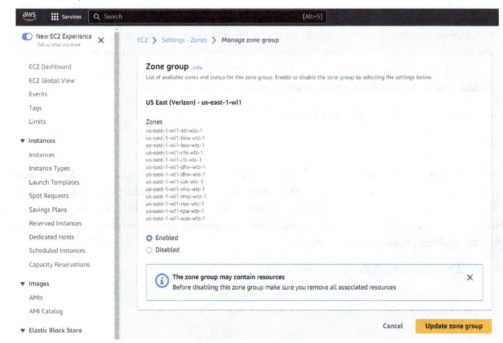

Figure 7.2 – Enabling AWS Wavelength Zones for us-east-1

Within the region you wish to enable them for, go to the EC2 dashboard and select **Zones**, scroll down to **Wavelength Zones**, click **Manage**, and then set the one(s) you want to see as **Enabled** (see *Figure 7.2*). Alternatively, you may use the `ModifyAvailabilityZoneGroup` API action using the CLI as follows:

```
export REGION=<your region goes here>

aws ec2 modify-availability-zone-group \
          --group-name ${REGION}-wl1 \
          --opt-in-status opted-in \
          --region $REGION

aws ec2 describe-availability-zones \
          --region $REGION \
          --query 'AvailabilityZones[*].ZoneName'
```

```
seahow@seahowdesk:~$ aws ec2 describe-availability-zones \
                     --region eu-west-2 \
                     --query 'AvailabilityZones[*].ZoneName'

[
    "eu-west-2a",
    "eu-west-2b",
    "eu-west-2c",
    "eu-west-2-wl1-lon-wlz-1",
    "eu-west-2-wl1-man-wlz-1"
]
```

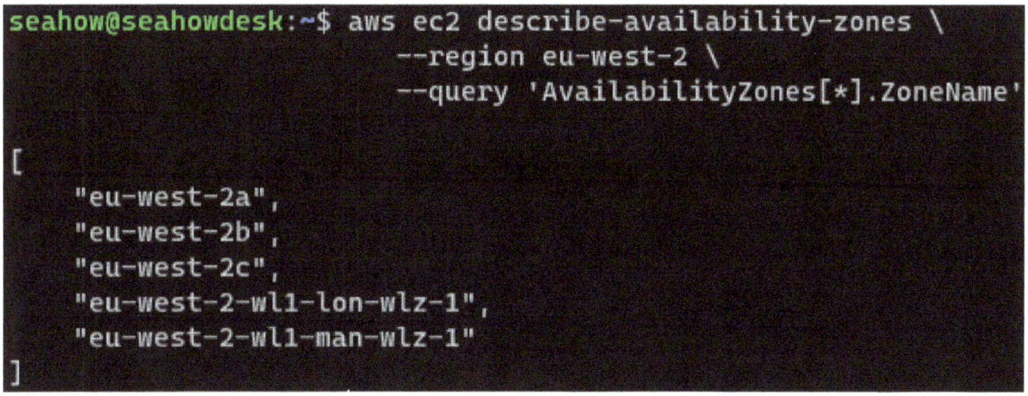

Figure 7.3 – describe-availability-zones CLI command output with AWS Wavelength enabled

Once the Wavelength Zone(s) are enabled, you can now create subnets within them just like you would for any other availability zone in your VPC:

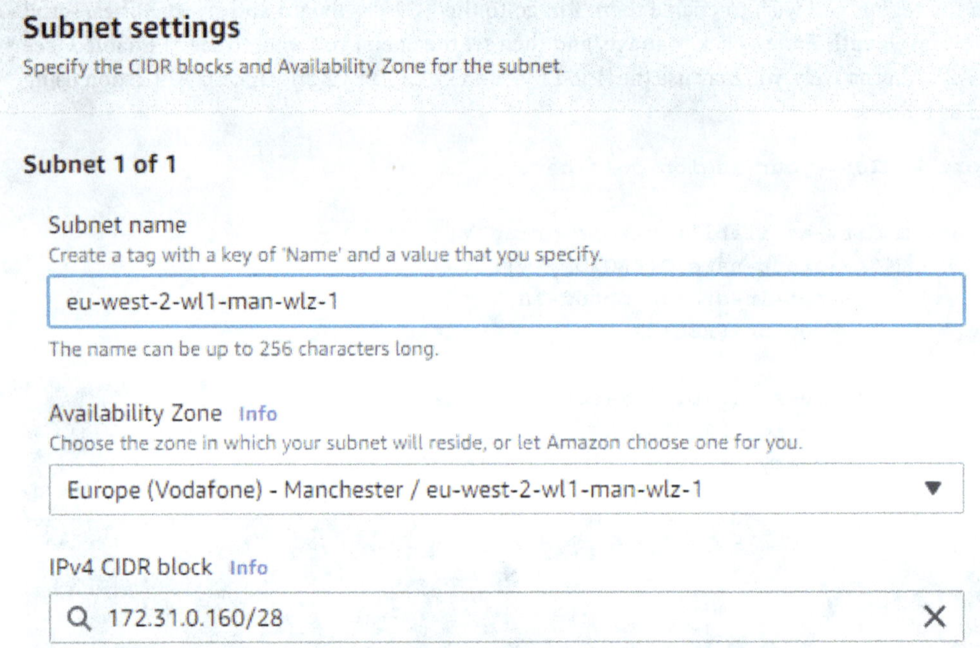

Subnet settings

Specify the CIDR blocks and Availability Zone for the subnet.

Subnet 1 of 1

Subnet name
Create a tag with a key of 'Name' and a value that you specify.

```
eu-west-2-wl1-man-wlz-1
```

The name can be up to 256 characters long.

Availability Zone   Info
Choose the zone in which your subnet will reside, or let Amazon choose one for you.

```
Europe (Vodafone) - Manchester / eu-west-2-wl1-man-wlz-1          ▼
```

IPv4 CIDR block   Info

```
Q   172.31.0.160/28                                              X
```

Figure 7.4 – Creating a subnet for a Wavelength Zone

Carrier gateways

AWS Wavelength Zones use a special construct similar to an internet gateway called a carrier gateway. This is how traffic will route to/from the MNO's network to EC2 instances or containers you attach to the subnets in your AWS Wavelength Zone:

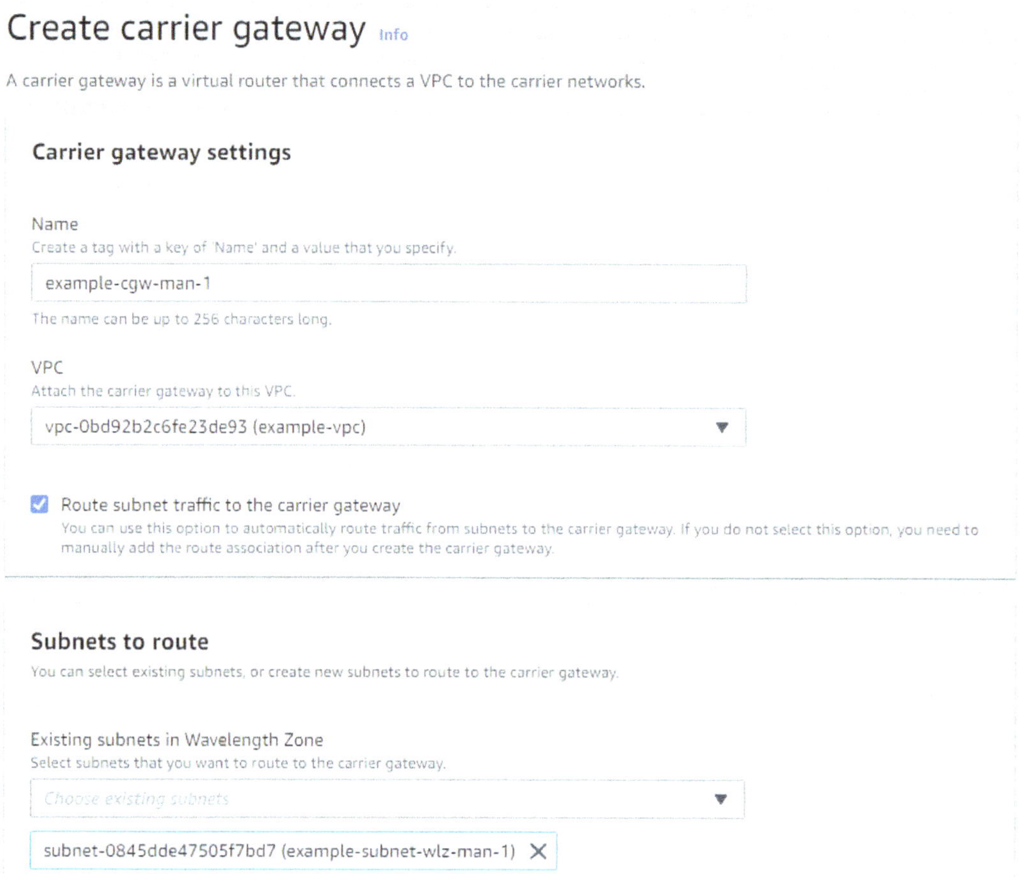

Create carrier gateway Info

A carrier gateway is a virtual router that connects a VPC to the carrier networks.

Carrier gateway settings

Name
Create a tag with a key of 'Name' and a value that you specify.

> example-cgw-man-1

The name can be up to 256 characters long.

VPC
Attach the carrier gateway to this VPC.

> vpc-0bd92b2c6fe23de93 (example-vpc)   ▼

☑ Route subnet traffic to the carrier gateway
     You can use this option to automatically route traffic from subnets to the carrier gateway. If you do not select this option, you need to
     manually add the route association after you create the carrier gateway.

Subnets to route

You can select existing subnets, or create new subnets to route to the carrier gateway.

Existing subnets in Wavelength Zone
Select subnets that you want to route to the carrier gateway.

> Choose existing subnets   ▼

> subnet-0845dde47505f7bd7 (example-subnet-wlz-man-1) ✕

Figure 7.5 – Creating a carrier gateway for an AWS Wavelength Zone

Routing tables

When using Wavelength Zones, it is best to have a different route table in the VPC from the one the public subnets in the parent region are attached to. This is so that only resources in the Wavelength Zones have their non-VPC traffic routed directly onto the MNO's network instead of out of the usual internet gateway:

Figure 7.6 – Route table defaulting to a carrier gateway

Once a carrier gateway exists, set the default route ($0.0.0.0/0$) to point to it within the relevant route table(s). Note that there is only ever one carrier gateway per VPC:

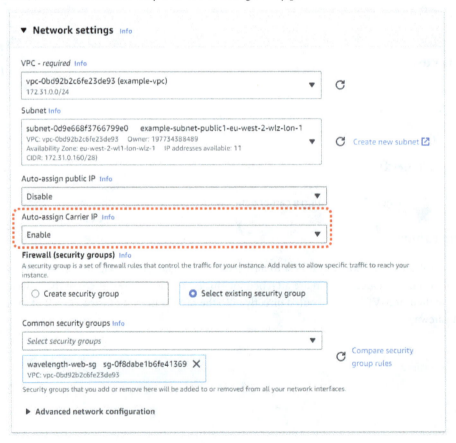

Figure 7.7 – Deploying an EC2 instance with a carrier IP

Once the subnets and routes are configured, you can begin deploying compute resources into a Wavelength Zone. Unlike EC2 instances in standard AZs, instances deployed into one of these special zones can assign a carrier IP at deploy time. It is essentially the same thing as a public IP in a region, only it is specifically drawn from the MNO's IP space. That's how traffic goes directly to them. Note that you cannot assign standard public IPs to resources in Wavelength Zones.

It is important to understand that only subscribers on a given MNO's network can access a carrier IP. For example, in the next diagram, only Verizon customers can reach 141.195.196.252 or 141.195.196.95. Subscribers on other networks or devices on the general internet will get timeouts if they attempt to reach those IPs:

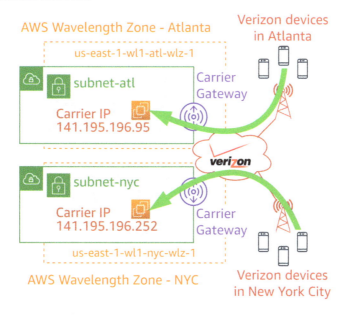

Figure 7.8 – Ingress from Verizon network to AWS Wavelength Zones

Differences between 5G and 4G/LTE clients

Both types of devices on an MNO's network can connect to resources in an AWS Wavelength Zone. However, the latency is dramatically different. Recall from *Chapter 3* the architectural differences between 5G and 4G/LTE topologies.

A 5G connection can leave an MNO's network via one of many distributed points known as the **User Plane Function** (**UPF**). MNOs can use 5G network slicing to ensure latencies as low as a single millisecond from the device to the carrier gateway.

By comparison, a 4G/LTE device must backhaul to a central point of egress from the MNO's network – the **Packet Gateway** (**PGW**). Further, the MNO's ability to ensure a high **quality of service** (**QoS**)

is lessened in the absence of network slicing. It is not uncommon for a device that can reach a carrier gateway in 1 ms over 5G requires 30 ms when switched to 4G/LTE mode.

If a compute resource on a public subnet within an AWS Wavelength Zone needs to initiate a connection out to the internet, it will do so using the carrier gateway and get to the internet via whatever path that MNO uses for internet access.

Internet access

A container or instance in an AWS Wavelength Zone can use a carrier gateway to connect to servers on the internet via the MNO's network:

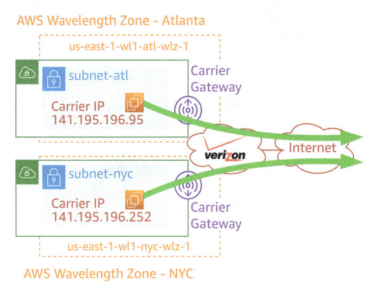

Figure 7.9 – Egress to the internet via carrier gateways

When doing this, outbound TCP connections from AWS to the internet function normally. However, UDP connections are blocked over this path.

Application Load Balancer

When an **Elastic Load Balancer** (**ELB**) is set up in a Wavelength Zone, it is done in much the same way as it is in any standard availability zone, with three differences. First, unlike **Application Load Balancers** (**ALBs**) in a standard region, only one subnet must be selected. Second, only the Layer 7 ALB variant is available. Layer 4 or **Network Load Balancers** (**NLBs**) cannot be deployed to a Wavelength Zone. Lastly, the listener is going to be on the carrier's network and use a carrier IP:

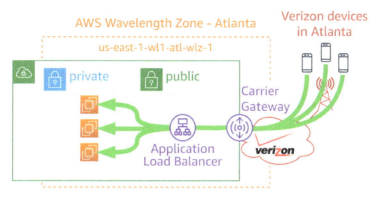

Figure 7.10 – ALB in an AWS Wavelength Zone

Special note on the geographical availability of ALBs

At the time of writing, ALBs can only be deployed to the 19 AWS Wavelength Zones in the US. If you attempt to deploy an ALB to an unsupported zone, you will receive an error. Customers in other geographies must use AWS Route53, third-party appliances from the AWS Marketplace, self-configured HAProxy, or a similar alternative.

Amazon Route53 for load balancing

For non-HTTP(s) applications or AWS Wavelength Zones outside the US, it is possible to use Amazon Route53 weighted routing policies for load balancing (see *Chapter 6*). However, unlike using it in standard regions or AWS Local Zones, it is not possible to use the health check feature in the normal way. This is because the Amazon Route53 health checkers do not have access to the carrier IPs on any MNO's 5G fabric:

☑	london-hc		CloudWatch alarm: awsec2-i-022952f3659813a05-GreaterThanOrEqualToThre

15 minutes ago

Info	Monitoring	Alarms	Tags	**Health checkers**	Latency

● View current status   View last failed check   🗘 Refresh

Health checker region	Health checker IP	Last checked	Status
US West (N. California)	15.177.10.18	Aug 18, 2023 12:30:58 PM UTC	Success: 1 datapoint
US West (N. California)	15.177.14.18	Aug 18, 2023 12:30:57 PM UTC	Success: 1 datapoint
US West (Oregon)	15.177.22.18	Aug 18, 2023 12:30:59 PM UTC	Success: 1 datapoint
US West (Oregon)	15.177.18.18	Aug 18, 2023 12:30:58 PM UTC	Success: 1 datapoint
EU (Ireland)	15.177.34.18	Aug 18, 2023 12:31:00 PM UTC	Success: 1 datapoint

Figure 7.11 – Using an EC2 instance's status check to determine health

Instead, you can create an Amazon CloudWatch alarm that monitors something the AWS control plane is already aware of. For instance, you could set the alarm to watch an EC2 instance's status checks. In turn, you configure the Amazon Route53 health check to monitor this alarm. The preceding figure shows an example of this configuration.

Edge Discovery Service (EDS)

One of the key benefits of building an application that runs directly on an MNO's 5G edge is the ability to take advantage of network slicing. This allows us to maintain QoS parameters from the mobile device to a server in an MEC architecture. However, this only works if the traffic never leaves the MNO's 5G fabric. The moment traffic leaves the UPF and enters the general internet, the MNO loses control of connection quality.

This is no big deal for an application that lives only in one AWS Wavelength Zone. The load-balancing mechanisms discussed so far can handle that. But in cases where our app spans multiple geographies, how do we make sure traffic goes from a given mobile device to the closest carrier gateway? Because the IP address of a mobile device (also known as a UE Identity) isn't routable on the internet, IP Anycast services such as AWS Global Accelerator can't help us. Nor can **Global Server Load Balancing** (GSLB) mechanisms such as Amazon Route53.

Client-side triangulation

One approach would be to do ping triangulation from within the application on the mobile device, as shown in the following diagram:

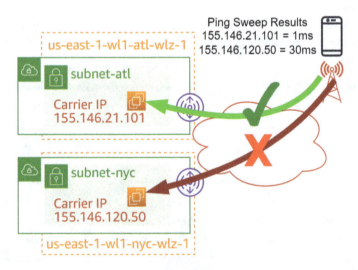

Figure 7.12 – Ping triangulation from a mobile device

This can be tricky because it requires a mechanism to keep the list of carrier IPs updated on mobile devices. Customers who implement this typically use a managed GraphQL service such as AWS AppSync. GraphQL is an alternative to REST APIs. It supports a publish/subscribe model as opposed to the constant polling from all mobile devices involved that would be needed with REST.

> **Law of physics versus law of the land**
>
> The device triangulation method necessarily finds the best zone in terms of performance – but an application with compliance requirements to communicate only with servers in a certain country or state might be better served using another method.

AWS Cloud Map

Another approach is to maintain a central directory of carrier IPs that mobile applications can use to look up the closest carrier IP given their current GPS coordinates. This can be done using AWS Cloud Map. This is a managed cloud resource discovery service that is integrated with other AWS services:

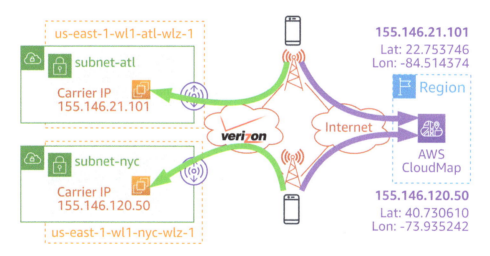

Figure 7.13 – Using AWS Cloud Map to discover carrier IPs based on GPS coordinates

AWS Cloud Map allows you to create namespaces that dynamically track the state and location of constructs such as EC2 instances or containers in ECS/EKS. Because it is possible to attach arbitrary attributes to such things via tags, GPS coordinates for a given carrier IP can be retrieved by the mobile device. At that point, it would be up to the application on the device to calculate the closest carrier IP given the device's location. In addition, the application needs permission to retrieve its location from the device – something that not all users are comfortable with.

Carrier-developed EDS

Another method is to leverage the fact that the MNO knows the physical location of all devices on its 5G fabric. Carriers such as Verizon have developed their own service discovery APIs that they make available on their 5G fabric. Mobile devices can query Verizon's EDS service, which already knows both the physical location of the mobile device's UE Identity and all carrier IPs it is servicing:

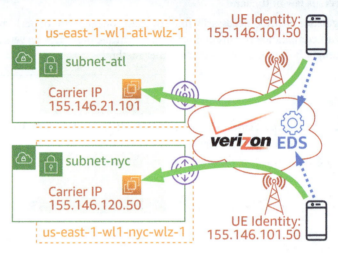

Figure 7.14 – Dynamic edge discovery with Verizon EDS

This approach simplifies the implementation on the mobile application by removing the requirement for permission to query the device's GPS. It also eliminates the need to calculate coordinates and figure out which carrier IP is best. However, this comes at the cost of increased management overhead associated with keeping your application's services registered in service profiles in Verizon's EDS.

Security considerations

Carrier-developed EDS systems require client devices to be in possession of API keys that grant access to determine the location based on the client IP address/UE Identity. Depending upon the application's architecture, granting these keys to thousands of devices could represent a vulnerability. Some customers, therefore, implement a caching tier on a central AWS resource (such as a container in the parent region). Only this central caching tier interacts with the EDS system, providing an opportunity for the application owner to directly inspect calls and respond to anomalous behavior.

Accommodating device handoff

It is important that applications that span multiple geographies take into consideration the possibility that the closest AWS Wavelength Zone might change as a client device moves. A car with a 5G modem in it doing **vehicle-to-everything (V2X)** could easily drive from Boston to Philadelphia via New York – necessitating a switch between those three zones as it moves. Regardless of the discovery mechanism in use – be it triangulation, carrier-developed EDS, or self-developed EDS – the application should periodically re-evaluate the best carrier IP to use.

Extending a VPC into AWS Wavelength

When extending a VPC from a parent region into AWS Wavelength Zones, many of the considerations seen with AWS Outposts and AWS Local Zones remain the same.

Communication between AWS Wavelength Zones

Just as with AWS Outposts and AWS Local Zones, it is not possible to route traffic from one zone to another via the parent region within the same VPC. As shown in the next diagram, multiple VPCs must be configured, and a mechanism such as AWS Transit Gateway should be used to route between them within the parent region:

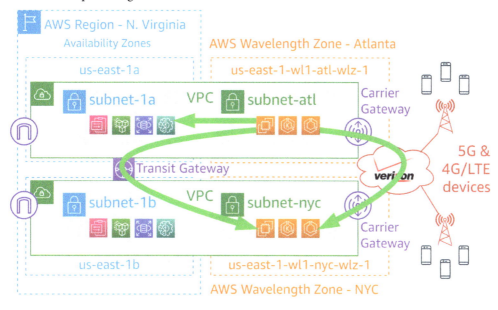

Figure 7.15 – Communication paths to other VPC subnets

It is, however, possible for resources in one zone to talk to resources in another over the MNO's network using their carrier IPs.

> **Maximum Transmission Unit (MTU) values for AWS Wavelength:**
>
> 9,001 bytes between EC2 instances within the same AWS Wavelength Zone
>
> 1,500 bytes across the carrier gateway
>
> 1,468 bytes between an AWS Direct Connect instance and an AWS Wavelength Zone
>
> 1,300 bytes between an EC2 instance in an AWS Wavelength Zone and an EC2 instance in a standard region

Communicating to AWS endpoints

In an AWS Region, you don't need to think very hard when communicating with AWS services. The communication with them will happen via your internet gateway or **network address translation (NAT)** gateway as applicable. However, on a public subnet in an AWS Wavelength Zone, your route table will have 0.0.0.0/0 pointing to the carrier gateway:

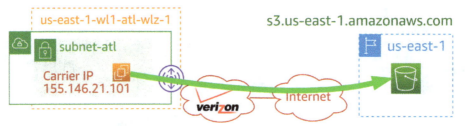

Figure 7.16 – Communicating with public endpoints over an MNO's network

This means, by default, any connections open to AWS services such as Amazon S3 will traverse the MNO's network to get to the internet – only to turn around and come back over the public internet to the parent region where the public endpoint lives:

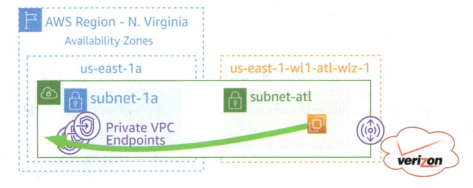

Figure 7.17 – Using AWS PrivateLink endpoints in the parent region

It is, therefore, recommended to use private endpoints for any AWS services your application needs to communicate with.

> **Using private VPC endpoints in AWS Wavelength Zones**
>
> VPC interface endpoints must be attached to subnets in the parent region. They cannot be attached to subnets in an AWS Wavelength Zone.

This also needs to be considered for any services your mobile application itself directly uses. It is common for mobile devices to retrieve files from Amazon S3:

Figure 7.18 – Mobile device communicating with Amazon S3 over the internet

This is fine in many circumstances. Just remember – once the connection leaves the MNO's mobile network for the open internet, the QoS drops to best effort:

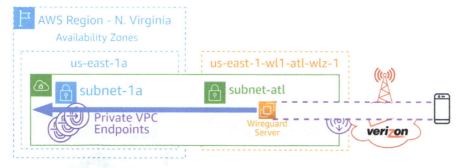

Figure 7.19 – WireGuard VPN used to access private endpoints

In cases where the QoS for these connections is important, consider implementing a lightweight VPN such as WireGuard to an EC2 instance inside an AWS Wavelength Zone. This allows you to route traffic to AWS PrivateLink endpoints in the VPC by impersonating the private IP of the VPN server.

While standard Linux-based VPN services such as Openswan can also perform this function, WireGuard has a much smaller footprint and makes better use of available compute resources. It is possible to push 500 Mbps over a WireGuard server on a t3.micro EC2 instance.

This is largely due to its simplicity. WireGuard consists of less than 7,000 lines of code. By comparison, StrongSwan + XFRM (a common IPSec implementation) is over 400,000 lines – and OpenVPN + OpenSSL (a common SSL VPN implementation) is greater than 600,000.

Considerations such as this are important in the world of embedded devices.

Integrating AWS Wavelength with other services

Referring back to the section in *Chapter 3* about MEC, we covered how MNOs are building these capabilities out in their existing aggregation or regional **central offices** (**COs**). These are not gigantic data centers such as those a standard availability zone consists of.

This is why the service selection is limited to those that are most requested by customers for proven MEC use cases.

EC2 instances

AWS Wavelength Zones supports a limited number of Nitro-based Amazon EC2 instance types. The current list of instance types available in any given AWS Wavelength Zone can be queried with the following CLI command:

```
aws ec2 describe-instance-type-offerings \
--location-type "availability-zone" \
--filters Name=location,Values=eu-west-2-wl1-lon-wlz-1 \
--region eu-west-2 \
--query "InstanceTypeOfferings[*].[InstanceType]" \
--output text | sort
```

```
seahow@seahowdesk:~$ aws ec2 describe-instance-type-offerings \
                     --location-type "availability-zone" \
                     --filters Name=location,Values=eu-west-2-wl1-lon-wlz-1 \
                     --region eu-west-2 \
                     --query "InstanceTypeOfferings[*].[InstanceType]" \
                     --output text | sort
g4dn.2xlarge
r5.2xlarge
t3.medium
t3.xlarge
```

Figure 7.20 – Querying the EC2 instance types available in the AWS Wavelength Zone in London

Dev/test applications with smaller footprints can be launched by anyone on demand in AWS Wavelength. However, it is important to remember the capacity in any given location is managed by the MNO in whose network it lives. It is prudent to reach out to both your MNO and your AWS account team when planning larger deployments to ensure adequate capacity is provisioned and possibly reserved on your behalf.

> **Limits on dedicated EC2 resources**
>
> Dedicated Instances and Dedicated Hosts are not available in AWS Wavelength Zones.

Amazon ECS

ECS is available in AWS Wavelength in two forms – Standard ECS on EC2 and ECS Anywhere. However, ECS on AWS Fargate is not supported.

Applications that span multiple AWS Wavelength Zones are really intended to function in a hub-and-spoke model. While it is possible through the use of multiple VPCs and AWS Transit Gateway to enable communication between a container in, say, New York City and one in Atlanta, this is usually a bad idea. It defeats the entire purpose of moving resources to such a specific location to reduce latency.

Microservice-based applications designed for orchestrated containerized environments tend to form a mesh of dependencies. The orchestration layer often does things such as ensuring two instances of a given microservice container are kept on separate physical nodes for high availability reasons. However, they generally don't measure or consider the latency across the entire chain of dependencies as it pertains to the initial requester. In other words, if you deploy an ECS cluster and simply add all of your AWS Wavelength Zone subnets to the default capacity provider, you will end up with cross-zone communication paths that you don't want:

Subnets

Select the subnets where your tasks run. We recommend that you use three subnets for production.

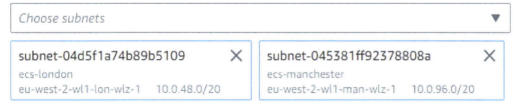

Default namespace - *optional*

Select the namespace to specify a group of services that make up your application. You can overwrite this value at the service level.

Figure 7.21 – The wrong way to deploy an ECS cluster with AWS Wavelength Zones

This is why it is ideal with ECS to maintain separate EC2 Auto Scaling Groups as distinct capacity providers across AWS Wavelength Zones. These capacity providers should also fall under separate namespaces in AWS Cloud Map. The next diagram shows this topology:

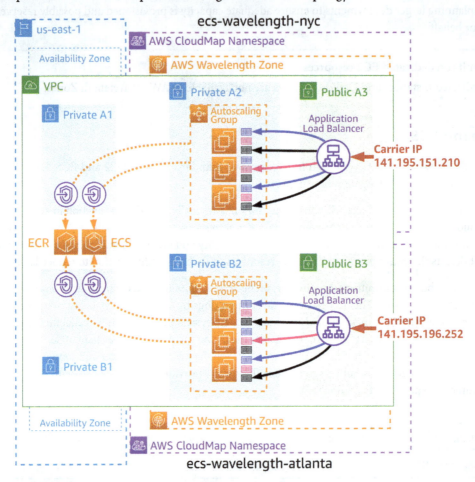

Figure 7.22 – ECS cluster spanning two AWS Wavelength Zones with separate namespaces

Amazon EKS

Distributed applications running on EKS are supported within AWS Wavelength Zones. However, there is only one supported deployment model. In the same way extended clusters work with AWS Outposts, the EKS control plane components must be deployed within the parent region. The next diagram shows an example of this setup

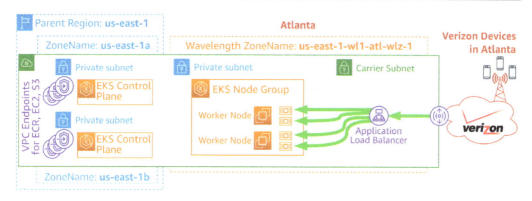

Figure 7.23 – Self-managed EKS node group in an AWS Wavelength Zone

Note that you cannot deploy managed EKS node groups into AWS Wavelength Zones; only self-managed node groups are supported. This means the worker nodes are part of an Auto Scaling Group, and they run the `/etc/eks/bootstrap.sh` script upon launch that tells them how to join the K8s cluster directly. This script is part of the EKS **Amazon Machine Image** (**AMI**) that AWS publishes. The latest version of this AMI can be found with the following CLI command:

```
aws ssm get-parameter \
--name /aws/service/eks/optimized-ami/1.17/amazon-linux-2/recommended/image_id \
--region eu-west-2 \
--output text \
--query 'Parameter.Value'
```

```
seahow@seahowdesk:~$ aws ssm get-parameter \
                --name /aws/service/eks/optimized-ami/1.17/amazon-linux-2/recommended/image_id \
                --region eu-west-2 \
                --output text \
                --query 'Parameter.Value'
ami-0cec489756f45eda0
```

Figure 7.24 – Obtaining the ID for the latest EKS optimized AMI in eu-west-2 from SSM

> **Self-managed node groups for EKS**
>
> For more information, see `Launching self-managed Amazon Linux 2 nodes` in the *Amazon EKS User Guide*. Also, read `Managing users or IAM roles` in the *Amazon EKS User Guide* for information on applying the `aws-auth` ConfigMap to your EKS cluster.

Service pricing

It is important to be aware of the pricing difference for AWS services when they are deployed to a Wavelength Zone versus a standard availability zone. The next figure shows an example of the difference for different EC2 instance types in us-east-1 (Northern Virginia region) and us-east-1-atl-1a (Atlanta Wavelength Zone):

Instance type ▽	On-Demand Linux pricing ▽	On-Demand us-east-1-atl-1a Linux pricing
t3.medium	0.0416 USD per Hour	0.052 USD per Hour
t3.xlarge	0.1664 USD per Hour	0.208 USD per Hour
g4dn.2xlarge	0.752 USD per Hour	1.015 USD per Hour

Figure 7.25 – On-demand pricing differences for AWS Wavelength Zones

As you can see, the costs for the instance types in question are anywhere from 25% to 35% more per hour in a Wavelength Zone.

Summary

The introduction of AWS Wavelength Zones marks a significant stride in AWS's efforts to seamlessly integrate cloud services with mobile edge computing, facilitating applications that require single-digit millisecond latencies. Customer needs are ever-evolving, and the push toward more immediate data processing, especially in the realm of 5G, has become apparent.

In this chapter, we discussed how mobile device connections are facilitated through carrier gateways, ensuring a smooth interaction between mobile networks and AWS services. Tools such as Route53 and Global Accelerator further optimize and direct the traffic, ensuring efficient routing. We also reviewed approaches to extending a VPC into an AWS Wavelength Zone, ensuring a seamless and secure environment for their applications.

Lastly, whether running EC2 instances, managing traffic with ALBs, or deploying applications with ECS or EKS, AWS ensures that users have a consistent and familiar experience.

In the next chapter, we will dive deep into edge computing services offered within the more than 450 **points of presence** (**PoPs**) AWS maintains globally.

Part 3:
Building Distributed Edge Architectures with AWS Edge Computing Services

Part Three dives into how you can use the AWS services introduced in *Part Two* to address common customer requirements. It also covers how these services integrate with other commonly used AWS services you are already familiar with. This will enable you to architect distributed edge applications that extend from the core of the AWS cloud to the farthest edge devices.

This part has the following chapters:

- *Chapter 8, Utilizing the Capabilities of the AWS Global Network at the Near Edge*
- *Chapter 9, Architecting for Disconnected Edge Computing Scenarios*
- *Chapter 10, Utilizing Public 5G Networks for Multi-Access Edge (MEC) Architectures*
- *Chapter 11, Addressing the Requirements of Immersive Experiences with AWS*

8

Utilizing the Capabilities of the AWS Global Network at the Near Edge

The AWS Global Network is the cornerstone of everything AWS does. It represents an expansive and cutting-edge infrastructure designed to offer unparalleled performance, reliability, and security. It's more than just a network; it's a testament to AWS's commitment to delivering a seamless cloud experience to businesses and end users across the world. By understanding and harnessing the full capabilities of this network, organizations can unlock new potential in application performance, global content delivery, and data transport strategies.

The AWS Global Network offers several near-edge services at the 450+ locations where it peers with the internet. In this chapter, we will cover the following topics:

- Overview of the AWS Global Network
- Processing at the near edge with Amazon CloudFront
- Leveraging IP Anycast with AWS Global Accelerator
- Using the AWS Global Network as a private WAN

Overview of the AWS Global Network

Networks like those run by AT&T, Comcast, NTT, Tata, or Zayo are known as Tier-1. This means they can reach the entire internet through settlement-free peering. These are the biggest players on the internet, with the most resources. Nevertheless, even Tier-1 ISPs oversubscribe their networks.

This means they sell more bandwidth to their customers than they can deliver – all at once, anyway. They rely on the fact that most customers don't hammer their full bandwidth allocation 24 hours a day. This is why your ISP probably gives you less upload than download speed, and why they implement data caps. They also use complex QoS mechanisms to deal with the inevitable periods of congestion that result.

Over the years, they have become good at predicting how much oversubscription they can get away with before losing customers to competitors. For context, an ISP that only oversubscribes 25:1 is considered a good one. Ratios as high as 100:1 are not uncommon.

AWS went online in 2006. In the early days, transit between AWS regions happened over Tier-1 networks. After a few years, it became clear that the needs of a cloud service provider were quite different than those of an ISP. Therefore, AWS built a private backbone between all of its regions and edge locations.

AWS Global Network

In contrast to the oversubscription seen with Tier-1 ISPs, AWS' backbone is overprovisioned. This means AWS maintains additional capacity above and beyond what it needs most of the time. Rather than rely on QoS to deal with congestion caused by unexpected bursts of traffic, AWS builds to prevent congestion from happening in the first place.

This allows AWS to ensure a deterministic level of performance. It also allows them to do things such as allow MTUs of 9,001 between regions or implement optimizations such as TCP termination from its edge POPs

This total control over all transit is the basis of services such as AWS Global Accelerator, AWS CloudFront, and AWS CloudWAN – all of which will be covered later in this chapter (see the *Processing at the near edge with Amazon CloudFront*, *Leveraging IP Anycast with AWS Global Accelerator*, and *Using the AWS global backbone as a private WAN* sections for more details).

Availability Zones (AZs)

To understand the full benefits of utilizing the AWS global backbone, we must first start at the foundational units that make up the AWS cloud – AZs.

CLOS (leaf/spine) topology

The physical network fabric in AZs is a fully Layer 3 CLOS architecture, also known as a leaf/spine design. To simulate Layer-2 adjacency between EC2 instances in different parts of an AZ, AWS uses its own in-house developed **Software-Defined Networking** (**SDN**) platform.

This platform's control and management planes work in tight coordination with ASICs on the network cards inside the physical servers hosting EC2 instances. These ASICs perform encapsulation and routing functions for the data plane at tremendous speed – something only custom-built hardware can do.

This means that within an AZ, the logical architecture of a customer's VPC and subnets is completely divorced from the physical fabric. At the same time, due to the innovations AWS has made with custom silicon, performance and security are increased compared to non-virtualized networks.

Nitro

In 2017, AWS launched a new generation of EC2 instance types powered by a platform called Nitro, but its roots go back to 2013 with the **Enhanced Network Adapter** (**ENA**). The system contains multiple components:

- A hypervisor that is a heavily customized version of KVM

- A **Trusted Platform Module** (**TPM**) on the motherboard

- A separate security chip that does things such as secure the BIOS code

- Special I/O cards for storage and networking functions

It is these Nitro networking cards that we will focus on here. They are the foundation of network performance and security in AWS:

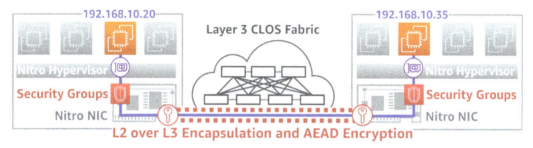

Figure 8.1 – Nitro NICs performing routing, encapsulation, and security functions in hardware

The preceding figure illustrates some of the essential functions the Nitro NICs perform for two EC2 instances in different parts of an AZ that want to communicate. In this example, both instances are a part of the same VPC, and on the same subnet (192.168.10.0/24).

From the perspective of the operating system inside both instances, they are on the same Layer 2 broadcast domain. When 192.168.10.20 opens a connection to .35, it issues an ARP broadcast like it normally would to discover the MAC address of .35. However, unlike a "real" Ethernet network, that broadcast never hits the network. The Nitro NIC intercepts it, performs an authenticated database lookup, and responds to it directly. All Layer 2 operations are managed in this way. That is why Layer 2 attacks such as poison ARP are non-sequiturs in AWS.

The next thing you will notice is the security group enforcement happening on the Nitro NICs. Security groups are stateful firewalls and they have no throughput limit, apart from whatever limit the EC2 instance type itself has. When the traffic leaves the physical server, the Nitro NIC further secures it using **Authenticated Encryption with Associated Data (AEAD)** algorithms, with 256-bit encryption.

Because all of this work is offloaded to specialized ASICs on the Nitro NICs, none of it imparts a latency or throughput penalty to the network flows.

AWS regional edge

When you create a VPC and attach an internet gateway or NAT gateway, the AZs in the associated region need a mechanism to connect to the internet or other regions:

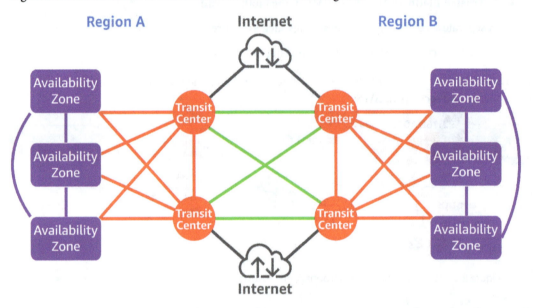

Figure 8.2 – AWS regional transit centers

This is why, in addition to the direct interconnects AZs have with each other, each region also has two independent, fully redundant **Transit Centers (TCs)**. TCs are how connections leaving an AZ reach another AWS region, access the internet, or communicate with AWS Direct Connect locations:

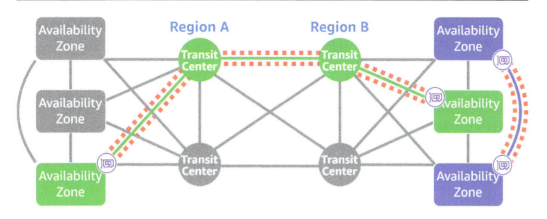

Figure 8.3 – Paths between ENIs in different locations

These TCs natively interconnect with AWS peers and **Internet Exchanges** (**IXs**) at anywhere from 100 to 600 Gbps (400 Gbps is the standard). To get an idea of the scale of the AWS Global Network's interconnections, navigate to the `PeeringDB entry for Amazon's AS`:

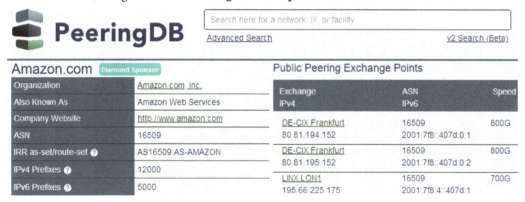

Figure 8.4 – PeeringDB entry for AS16509 (AMAZON)

Here, anyone can peruse the ~300 public exchange points and ~160 IXs AS 16509 is peered with, and at what speed.

Processing at the near edge with Amazon CloudFront

Amazon CloudFront is a worldwide **Content Distribution Network** (**CDN**), similar to Akamai or CloudFlare. CDNs are global (or sometimes regional) networks of proxy servers that cache copies of content closer to where user requests are coming from. The idea is that the cost of maintaining **Points-Of-Presence** (**POPs**) around the world is worth it because of the improved user experience due to lower latencies and/or the bandwidth saved by not transmitting the same large object across the internet over and over.

Fundamental to all CDNs is the ability to dynamically swap out the IP address of a server with one closer to the user each time a request comes in.

Content distribution

Amazon CloudFront uses a construct known as distributions to establish and govern this behavior. When a new distribution is created, it is pointed at an origin – this is the server or service that holds the content we want to cache in our POPs. Amazon CloudFront supports several types of origins – including S3 buckets, MediaStore containers/MediaPackage channels in Elemental, application load balancers, AWS Lambda functions, EC2 instances, and even servers that aren't part of AWS at all (custom origins):

Figure 8.5 – Content caching at the near edge

Consider a situation where we set up a new CloudFront distribution for an S3 bucket located at `https://mybucket.s3.eu-west-2.amazonaws.com/`. The first time a request comes in from a client in Helsinki for `s3://mybucket/image1.jpg`, the file will be retrieved by an edge POP in Helsinki from the London region (eu-west-2) and put into that edge POPs cache. This is known as a cache miss. 2 hours later, someone in Vaasa (close to Helsinki) requests `image1.jpg`. Because they are routed through the edge POP in Vaasa, they will retrieve the image directly from that POP's cache rather than having to go back to the London region. This is known as a cache hit. The ratio of hits to misses is called the cache hit ratio. Performance is best with a high ratio.

An object remains in the cache until it is expired, at which point the edge POP will delete it entirely from its cache (if no one has requested it in a long enough period) or fetch a fresh copy. This happens if it is still popular but has exceeded a designated **time-to-live** (**TTL**).

You don't want to cache everything. A good example would be files such as `index.php` or `cart.aspx` – these are dynamically generated. If you cache a shopping cart page, it will never update between users and you will never be able to add things to your basket. However, the GIF or JPG files for the products listed on these pages should be cached. It is not uncommon for a web page to consist of mixed elements like this – some static, some dynamic. Therefore, it is typical that some content be set to cache and other content to never cache. These rules can be set as behaviors of the distribution.

HTTP request headers and URL query strings

When a client on the internet uses their browser to make a request for a URL to a server, more information that the URL itself is passed in the form of HTTP request headers. This includes information such as the name and version of the web browser (user-agent), the language the client prefers (accept-language), and the DNS name of the server it wants to access (host). These fields are often relied upon by applications or web servers to operate properly.

For example, the (host) header field has been relied upon by web hosting companies for over 20 years – it allows them to host dozens or hundreds of websites on a single public IP address. Before including the (host) header became standard, web servers had to use a unique public IP for each website.

Unfortunately, a server is completely reliant upon the client to provide accurate information in these fields. Nothing is stopping a client from claiming to be a different type of device than it is (purposely or mistakenly). While this isn't a crime, it can break server-side mechanisms that do things such as hand out one video resolution for mobile devices and another for desktops.

With Amazon CloudFront, these request headers can be overwritten at the edge so that when a request comes into the origin server, it sees whatever headers, URLs, and query strings the edge wants it to see – not what the request contained.

Let's imagine a website that has grown over the years, such that JPGs in its `/images` folder are referred to in different ways by different parts of the site. One page might point to `/images/myimage.jpg`, while another might point to the same object as `/images/myimage.jpg?user=bob`. Those are going to be seen as different objects by a CDN such as Amazon CloudFront. This means they take up twice as much space in the edge cache and result in a lower overall cache hit rate.

For a moment, imagine that `myimage.jpg` isn't a flat file – instead, it is dynamically generated by a plugin on each request. In this case, the application relies on the (user-agent) header field to know which version of myimage.jpg to return. Should it return the high-resolution one or the low-resolution one? Maybe it relies on the (accept-language) field to know if it should hand out the one with French text in the Latin alphabet or the one with Bulgarian in Cyrillic. Unlike the situation with query strings, we have the opposite problem – the French version of myimage.jpg will cache as if it were the same object as the Bulgarian one. This is a problem because the first one that's returned will always be returned – sometimes inaccurately.

These are both examples of situations where you might want to alter the URL query string to force a caching behavior you prefer.

Amazon CloudFront special request headers

In *Chapter 2*, we discussed the mechanisms **Global Server Load Balancing (GSLB)** systems such as Amazon Route53 use to determine the geographic location of a client's IP address on the internet. Amazon CloudFront makes use of these same facilities to determine which edge location is closest to a given user so that it can steer them toward a cache containing the object they requested. It makes sense to use this data to populate additional header fields and pass them along for decision-making whenever a viewer makes a request:

Category	Fields	Examples
Country	Name	United States
Region	Region, Region-Name	TX, Texas
Locality	City, Postal-Code, Metro-Code	Dallas, 75001, 214
Coordinates	Latitude, Longitude	32.779167, -96.808891
Time	Time-Zone	America/Chicago
Device Type	Is-Mobile	True
	Is-Desktop	False
	Is-Tablet	True
	Is-IOS	True
	Is-Android	False
	Is-SmartTV	False

Figure 8.6 – Amazon CloudFront viewer headers for an iPad using 5G in Dallas, TX

These headers can be quickly enabled under the **Behaviors** section of an Amazon CloudFront distribution.

These headers make it easy for your distribution to do things such as the following:

- Retrieve an image or video that is appropriately sized for the client browser without needing to maintain a matrix of what user-agent equals what kind of device

- Respond with regionalized content that isn't reliant upon the client headers – just because someone has the language set to English doesn't mean they are in England

- Comply with regulations that say a website operator must make every effort to ensure certain content is only served within a given region, country, state, or municipality

Of course, the same problems GSLB faces are present here. Someone connecting to a VPN service in a certain region will appear to be in that region.

Regional edge caches (RECs)

RECs represent a second tier of caching that isn't as close as an edge location, but not as far as the origin. RECs are larger than edge POPs and therefore hold objects longer before being removed from the cache. This is useful in situations where a piece of content, say an MP4 video file for a training course, was heavily frequented by your users for a few days, then the activity dropped off... until a long 4-day weekend was over, at which point it began receiving requests again at 8 A.M. the first day back. Or maybe it's an image for a product being sold on an e-commerce site that doesn't get referred to often enough to be populated globally in all edge locations but does get hit enough to stay in the RECs consistently:

Figure 8.7 – Multiple stages of caching to limit the impact of misses

Amazon CloudFront Embedded POPs

In addition to the 450+ edge POPs that AWS maintains around the world, there are additional dedicated POPs positioned inside the networks of many ISPs/broadband providers around the world. These are known as Amazon CloudFront Embedded POPs, and they are based on AWS Outposts Rack or AWS Outposts Server (depending on the circumstances).

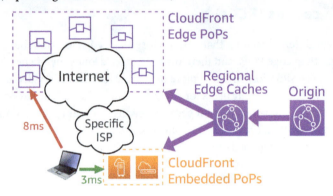

Figure 8.8 – Amazon CloudFront Embedded POPs

Large content providers such as NetFlix, HBO Max, and Amazon Prime use Amazon CloudFront to stream video to homes around the globe. These entities often negotiate with large ISPs to install these dedicated POPs into their networks for the explicit purpose of servicing, say, HBO Max viewers who are also Comcast customers. This preserves the ISP's bandwidth at their peering points and ensures a good user experience for the customers of the content provider. Over 220 of these were deployed in 2022 alone.

HTTP/3 and QUIC

Recall *Chapter 2*, when we covered the benefits of HTTP/3 and QUIC over TCP-based HTTP/2 and HTTP/1.1. Those benefits can be realized at the click of a button in an Amazon CloudFront distribution:

Supported HTTP versions

Add support for additional HTTP versions. HTTP/1.0 and HTTP/1.1 are supported by default.

 HTTP/2

☑ HTTP/3

Figure 8.9 – Enabling HTTP/3 for an Amazon CloudFront distribution

AWS has implemented HTTP/3 and QUIC on the Amazon CloudFront servers found in all edge POPs. Because those servers are full proxies between the viewer and the origin, no changes need to be made to the origin for this to work. Take a look at the following figure for an illustration of this:

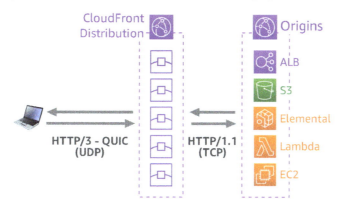

Figure 8.10 – HTTP/3 – QUIC distribution with no application changes

AWS Lambda@Edge

As we discussed earlier, there are times when the request headers Amazon CloudFront passes to the origin don't include enough information to make an appropriate decision about which object to return – or if we should respond at all. Sometimes, we need additional logic to make those determinations.

Let's imagine we have a type of content subject to legal restrictions that say we can only serve it to requesters not located in a certain country. We can easily geo-restrict this object based on the Amazon CloudFront geolocation headers – per the documentation, these are 99.8% accurate. However, they don't tell us if someone is using a VPN provider to *appear* as an IP in the correct location. As part of a mitigation strategy for this, we could cross-check the IP against a list of known VPN provider network prefixes.

An AWS Lambda function is ideal for something like this. AWS Lambda@Edge is just a version of AWS Lambda that runs in the RECs to be closer to the edge:

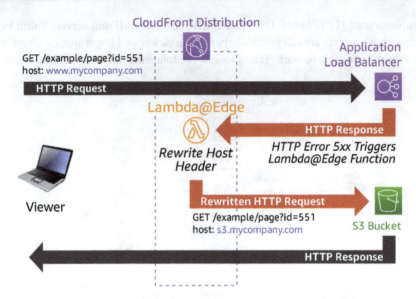

Figure 8.11 – Using AWS Lambda@Edge to alter HTTP response when errors occur

The preceding figure shows an example of using AWS Lambda@Edge to intercept an HTTP status code 500 error that an application behind an ALB generated. Instead of forwarding this raw error message to the viewer, this function is rewriting a new HTTP request for a specific object in an S3 bucket, and returning that to the user instead. This is commonly done to provide more elegant feedback to the end user. This is an example of an AWS Lambda@Edge function that is tied to the origin's response. That means it is triggered by, and works with, the response data from the ALB – as opposed to the inbound request data heading toward the origin.

The following figure shows the four different stages an AWS Lambda@Edge function can be associated with:

Function associations - *optional* Info

Choose an edge function to associate with this cache behavior, and the CloudFront event that invokes the function.

	Function type	Function ARN / Name
Viewer request	Lambda@Edge ▼	arn:aws:lambda:eu-west-2:457137(
Viewer response	Lambda@Edge ▼	arn:aws:lambda:eu-west-2:457137(
Origin request	Lambda@Edge ▼	arn:aws:lambda:eu-west-2:457137(
Origin response	Lambda@Edge ▼	arn:aws:lambda:eu-west-2:457137(

Figure 8.12 – Association types available for AWS Lambda@Edge

Note that AWS Lambda@Edge functions are somewhat limited compared to their counterparts running in the core regions. For instance, while an AWS Lambda function in a region can run for 15 minutes, those tied to the origin request/response have a maximum duration of 30 seconds, while those tied to the viewer request/response top out at 5 seconds. They also support a much more limited set of runtimes than those in regions. Check the `Amazon CloudFront Developer Guide for more details`.

Amazon CloudFront functions

As noted previously, AWS Lambda@Edge functions execute inside the RECs. That's better than having to run them in the core regions – but what can we do inside the edge POPs themselves? That is where we must use Amazon CloudFront functions:

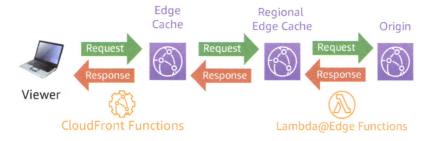

Figure 8.13 – Visualization of where Amazon CloudFront functions and AWS Lambda@Edge run

These are similar to Lambda@Edge and live within the same flow of a viewer request in the context of Amazon CloudFront. However, because they run at the edge POPs themselves, they are very close to the end users. They are also very fast – their startup times are sub-milliseconds. The trade-off is that the edge POPs don't have as many compute resources to go around, so they have to be lightweight – so lightweight that they are limited to a single millisecond for maximum duration. One way this is made possible is how they are written. Unlike the Node.js or Python runtimes supported by AWS Lambda@Edge, Amazon CloudFront functions are written in JavaScript:

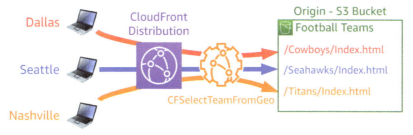

Figure 8.14 – An Amazon CloudFront function making decisions based on the user's geographic location

These are great for simple decision-making. Let's say that instead of outright georestricting content in our distribution, we want to steer users toward the landing page of the sports team in their area before they log into the site. This could be done using a function that looks at the CloudFront geolocation headers and modifies the request, as shown in the preceding figure.

Amazon CloudFront functions can only be associated with the viewer request/response stages. If you need to do something at the origin request or response, you will need to stick with AWS Lambda@Edge.

Leveraging IP Anycast with AWS Global Accelerator

In *Chapter 2*, we discussed how the primary causes of packet loss on the internet are congestion or throttling at the junction of two autonomous systems along the way:

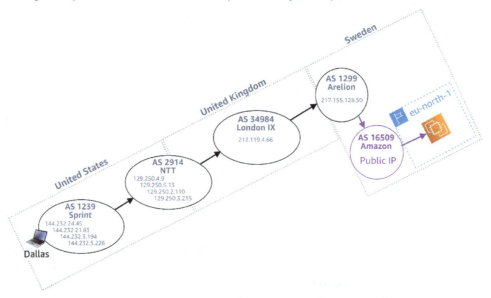

Figure 8.15 – A client in Dallas, USA, accessing an application in Stockholm, SE, via the public internet

Let's think about an application that runs on an EC2 instance in the AWS region in Stockholm, SE. The preceding figure provides an example of the path such a connection might take from a client in Dallas, USA.

As you can see, four different autonomous systems are traversed before the packets enter the AWS network in Stockholm. This is a lot of opportunity for congestion or QoS throttling mechanisms to cause a poor user experience.

Now, let's look at the same application, but this time with AWS Global Accelerator configured in front of it. Yes, the client in Dallas still has to physically get to Stockholm, so it's going to be a relatively high-latency connection:

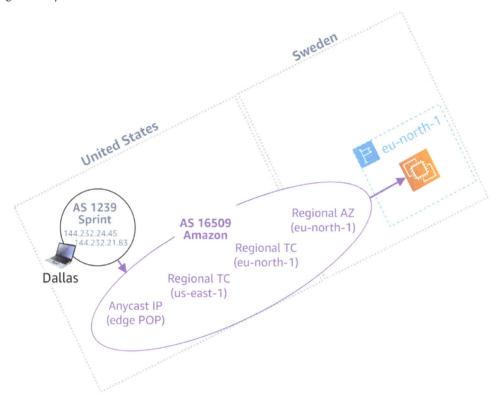

Figure 8.16 – A client in Dallas, USA accessing an application in Stockholm, SE, via AWS Global Accelerator

That said, take a look at the preceding figure to see the new path the connection takes. It enters the AWS network much earlier. This allows it to take a more direct path and one that isn't susceptible to congestion and throttling. Even though we've done nothing here to bring anything closer to the user, we have stabilized the path so that their experience is highly likely to improve – possibly enough that we don't have to distribute anything in the first place.

TCP termination

It is important to remember that while AWS Global Accelerator uses IP Anycast to steer clients into the nearest edge POP, it is much more than that. In a similar way to Amazon CloudFront, when customer connections enter an edge POP, they are terminated on a proxy server. This means that the three-way handshake – the initial setup of the TCP connection – happens much faster as it's confined to a shorter distance.

As you'll recall from our discussion in *Chapter 2*, latency in the TCP handshake interplays with the receive window to artificially limit the throughput of TCP connections. This situation is not made perfect by TCP termination, but it is significantly improved:

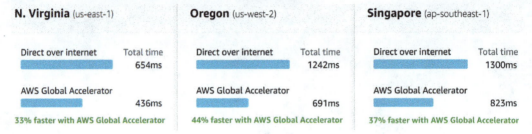

Figure 8.17 – AWS Global Accelerator Speed Comparison Tool from London, UK

To see the latency reduction and empirically test the transfer speed from different places in the world, visit the AWS Global Accelerator Speed Comparison Tool. The preceding figure shows an example of its output from a desktop in London, UK.

Endpoint groups

Endpoints can be thought of as similar to origins in Amazon CloudFront – they are the ultimate destination of a given connection:

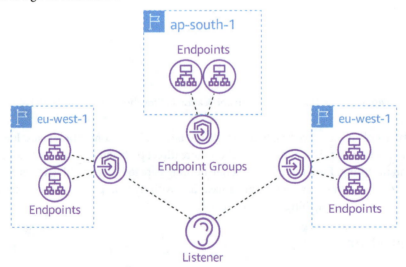

Figure 8.18 – AWS Global Accelerator overview of its logical hierarchy

The types of endpoints that are supported by AWS Global Accelerator include ALBs, NLBs, and EC2 instances in regions – not in AWS Wavelength or AWS Local Zones.

Preferential distribution of connections

New connections to a listener's Anycast IPs always enter the AWS network at the closest edge location. That's the whole idea of IP Anycast. From there, the default behavior is for a given connection to be sent to the closest region in which an endpoint group is configured for your listener.

Consider the scenario where you've configured a listener with two endpoint groups – one in Virginia and one in Tokyo. Each one has an EC2 instance endpoint in separate AZs in their respective region. Now, let's say a connection originates in Austin, TX, and enters the AWS backbone in Houston, TX. By default, the AWS Global Accelerator service in the Houston edge POP is going to send that connection to Virginia. Once it gets to Virginia, it will be routed to the EC2 instance in one of the AZs. A second connection from Dallas will go to Virginia as well, but it will get sent to the other AZ's EC2 instance in a round-robin fashion. This is just multiple tiers of load balancing like you're already used to with an ALB, for instance:

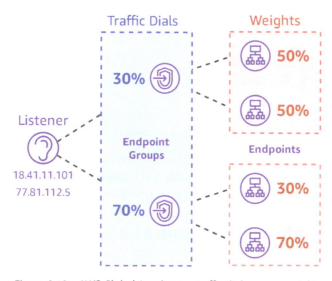

Figure 8.19 – AWS Global Accelerator traffic dials versus weights

However, there are situations where this behavior is not what we want. For example, you might want to set things up such that there is an active region and a standby region for a given application. Or maybe you wish to do blue/green swaps between application versions within a region. Perhaps you are chasing a difficult problem in a global application, and it's not clear if the problem is due to a connection's source or destination.

In these situations, it is possible to distribute connections arbitrarily between both regions and within them. This can be done at the regional/endpoint group level with traffic dials, or within a region at the endpoints using weights.

Traffic dials for endpoint groups/regions

By default, these are set to 100% for all endpoint groups attached to a listener. This results in the default behavior where connections always go to the closest region. When these values are altered, things can get a little tricky to understand, so let's walk through some examples:

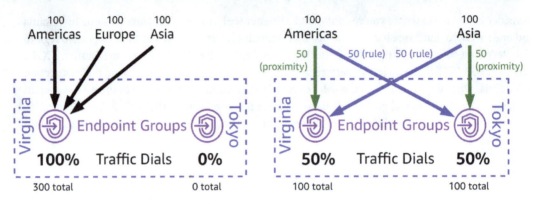

Figure 8.20 – AWS Global Accelerator traffic management across regions

First, let's take the simplest example of an active/standby setup. Here, we would set a traffic dial to 0 for the standby region and 100 for the active region. This is shown in the on the left of the preceding figure.

Second, let's say we set both regions to 50%. In this case, AWS Global Accelerator will ensure that 50% of connections are sent to both regions, regardless of where they originate from. The remaining connections will be allowed to follow the normal rules where they are sent to the closest region.

Using the AWS global backbone as a private WAN

Given the ubiquitous presence and high quality of the AWS Global Network, many customers have sought to build a private WAN infrastructure on top of it. This has been possible for some time through the use of third-party appliances from the AWS marketplace. Companies such as Aviatrix, Cisco, and Palo Alto Networks can build an overlay on top of EC2 instances running in different regions and use the AWS backbone as transport between them.

Until recently, however, there wasn't a native AWS service that could combine this with the power of AWS Direct Connect. Consider the following diagram. A customer has data centers in two different countries with AWS Direct Connect to their closest region and a gateway in the middle. It might make sense for some of the customer's traffic to never enter the AWS region at all and head straight for the other on-premises data center:

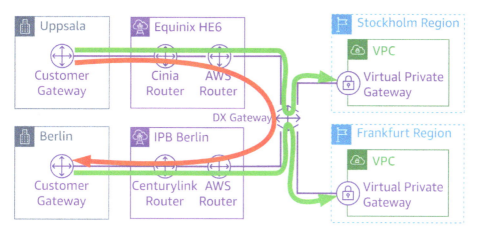

Figure 8.21 – AWS Direct Connect gateway paths

AWS Direct Connect SiteLink is a feature that enables routing from one **virtual interface (VIF)** connected to an AWS Direct Connect Gateway to another. It allows the red path shown in the preceding figure:

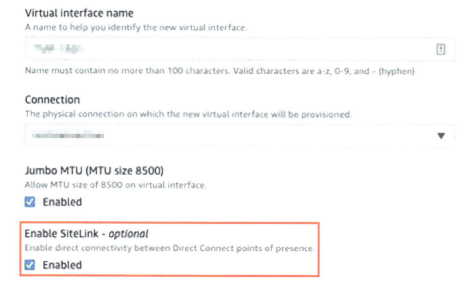

Figure 8.22 – Enabling AWS Direct Connect SiteLink on a VIF

AWS Cloud WAN

For relatively simple, static situations like what we've described so far, AWS Direct Connect SiteLink is sufficient. But let's say we want to build a SaaS offering on top of AWS that connects to physical elements at the edge. Some of these are in true data centers, with AWS Direct Connect links set up to their closest region. Others are remote edge locations that are more rugged, or even mobile. We could VPN from those locations to a region, but what if there are hundreds of such things? What if each one is owned by a different customer, and the traffic it sends across our private WAN needs to be compartmentalized from all other subscribers? Automating the provisioning of such a thing by spinning up EC2 instances and overlays could get out of hand pretty quickly.

This is where AWS Cloud WAN can help. Announced in July 2022, it is a service that lets you build, manage, and monitor a unified global network. Policies determine the network configuration and security rules. They can contain a construct called a segment that is isolated from other segments. These could map to each subscriber in the SaaS offering described previously. These policies also determine how these segments interact with VPCs, VPN clients, transit gateways, SD-WAN clients, and so on. Finally, they designate whether peering with other network constructs is allowed and in what manner.

Summary

Throughout this chapter, we dove deep into the unique architecture of the AWS Global Network. We started with foundational pillars such as the Nitro platform, which underpins the network's stability and performance, to advanced offerings such as Amazon CloudFront and AWS Global Accelerator that improve user experience no matter how far away from a region they are.

We also explored the transformative potential of processing at the near edge, a concept that has become central to modern content delivery and application deployment strategies. With tools such as AWS Lambda@Edge and Amazon CloudFront functions, AWS not only ensures optimal delivery speeds but also offers opportunities for real-time data processing and manipulation, empowering businesses to provide richer, more responsive user experiences. Moreover, by leveraging IP Anycast with AWS Global Accelerator and transforming the AWS infrastructure into a private WAN using offerings such as AWS Direct Connect SiteLink and AWS Cloud WAN, organizations are poised to craft global strategies that meet evolving business demands with agility and scalability.

In essence, the AWS Global Network is more than just a technical infrastructure; it's a dynamic ecosystem that's designed to adapt, evolve, and elevate. By fully understanding capabilities, businesses can achieve unparalleled global reach, performance, and resilience, ensuring that they stay ahead in an increasingly interconnected and digital world.

In the next chapter, we will explore patterns and anti-patterns when architecting for disconnected edge scenarios.

9

Architecting for Disconnected Edge Computing Scenarios

In this chapter, we will explore some of the most challenging edge computing scenarios – that is, those where connectivity back to an AWS region might be intermittent or non-existent. It is difficult to effectively run a globally distributed system when the endpoints aren't consistently reporting back. Even in such circumstances, AWS services allow customers to maintain operations while disconnected from the broader network.

As we proceed, we will build upon what we have learned in previous chapters to address such scenarios by joining AWS services with each other and with communication technologies such as SATCOM, 5G, and NB-IoT.

We will cover the following topics:

- Overview of DDIL
- SATCOM for DDIL
- Tactical edge
- Private 5G and DDIL

Overview of DDIL

In most situations involving cloud computing, consistent and reliable connectivity is table stakes. Some situations present unique challenges. This includes those where connections are prone to denial, disruption, intermittence, or limitation, collectively termed as **Denied, Disrupted, Intermittent, and Limited** (**DDIL**) connectivity. Let's take a closer look at these terms:

- **Denied connectivity**: This refers to environments where communication networks are unavailable or inaccessible. This could be due to a variety of reasons, including deliberate jamming, physical damage to infrastructure, or restrictive policies that block access to certain networks.

- **Disrupted connectivity**: This includes scenarios where established networks face intermittent disruptions due to various factors, such as environmental conditions, hardware malfunctions, or cyber-attacks. These disruptions can vary in scale and frequency, potentially leading to significant delays and loss of data.

- **Intermittent connectivity**: This is when connections are sporadic and not continuously available. This can often be seen in remote areas or during mobile communications where connections might be lost and regained periodically.

- **Limited connectivity**: These are settings where the available network resources are constrained, either due to bandwidth restrictions or limited coverage. The most typical examples are in remote areas where low-bandwidth satellite connections are all that is available.

Using AWS IoT services in DDIL scenarios

AWS has long been a leader in the IoT space. They offer a wide range of managed IoT services and customers have been operating IoT solutions on AWS at scale for over a decade. It is a very large and powerful toolbox, but at the end of the day, the challenge is right there in the name **IoT – Internet of Things**. What do you do when the **I** in IoT isn't a foregone conclusion?

Consider a situation where a few dozen IoT sensors and cameras are quickly set up in a remote location where even cellular service is not available. It would likely be prohibitively expensive to equip every sensor and camera with its own SATCOM terminal and associated service.

Even if we could afford to do that, the latency back to the region would be high and the throughput low – both of which would delay our ability to make decisions based on the data coming in from those sensors. This will typically be the case if we want to make **Machine Learning** (**ML**) inferences based on the combined data of several sensors.

For these reasons, and more, it makes sense to insert an aggregation point that those sensors and cameras can send their data to. At this point, a small ML model could make inferences based on multiple incoming streams. Audio, video, and other large pieces of data could be preprocessed – sending only the most critical and most optimized data over the expensive and slow SATCOM link. This aggregation point could also buffer incoming data in case the SATCOM terminal goes offline or becomes affected by environmental conditions such as rain fade.

The AWS Snow family of devices is ideal for providing the compute and storage needed for this purpose. Further, they can be ordered with a preconfigured AMI that runs AWS IoT Greengrass.

You will recall from *Chapter 1* that AWS IoT Greengrass can be visualized as a platform for running tiny versions of region-based AWS services you are already familiar with – including AWS SageMaker, AWS Lambda, AWS Kinesis Data Firehose, AWS Kinesis Video Streams, Amazon SNS, AWS Secrets Manager, AWS CloudWatch, and AWS Systems Manager.

The AWS Snow family as an IoT gateway

When running AWS IoT Greengrass on an AWS Snow family device, customers find the ability to host small ML models that have been pre-trained in the cloud particularly useful.

Here are some real-world examples:

- **Transcription/translation**: Passengers on trains cannot always hear the announcer (or hear them clearly). To improve accessibility, passengers can subscribe to an SMS service that will send transcribed text versions of what was said – translated into the passenger's native language if desired. Trains do have internet connections these days, but they tend to be slow and unreliable.

- **Mobile healthcare**: During the pandemic, customer demand for AWS Snow family devices exploded due to the need for portable on-site ML inferences for patient assessment and identification. It is straightforward to integrate a camera that streams to an AWS IoT Greengrass component, where facial recognition is performed. At the same time, a thermal camera sends an infrared map to a separate component that assesses whether the patient has a fever. Further, a third set of cameras observes the queue and another component estimates the queue waiting time based on object movement patterns. These interactions are then transferred to the main data store in the cloud as connectivity becomes available – or sometimes the device is simply returned to AWS for physical data loading and exchanged for a fresh one.

Let's assume that the relatively small amount of compute and storage available on an AWS Snowcone is sufficient for our use case. The following figure shows the high-level architecture we wish to achieve:

Figure 9.1 – AWS Snowcone physical connection via SATCOM

The first thing we will need to do is establish a communication channel over a SATCOM terminal to the internet.

Configuring the SATCOM terminal

In this example, we will use Inmarsat's L-band service (BGAN) to communicate with a geostationary satellite to reach the internet via an `AddValue Ranger 5000` terminal:

Figure 9.2 – A Ranger 5000 L-Band terminal (left) and using an IOS app to align it (right)

Recall from *Chapter 3* that geostationary satellites sit at a fixed point in the sky relative to a user on the ground. You will also recall from the same chapter that circular polarization is used in such situations, so our terminal's antenna is a flat square that needs to be pointed at whichever satellite in the constellation we have the best line of sight to.

There are multiple applications freely available for both iOS and Android that will assist in this process. See the preceding figure for an example of this. This particular terminal can also emit a tone while you are pointing the antenna to help fine-tune its positioning:

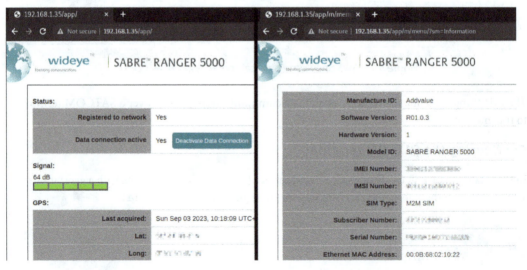

Figure 9.3 – Ranger 5000 terminal admin interface

Similar to the router an ISP might provide, the terminal has Ethernet ports that hand out DHCP addresses over to client machines. In this case, we will connect our laptop to Ethernet port 1 on the terminal and AWS Snowcone to Ethernet port 2. Before powering on the AWS Snowcone device, we must log in to the web interface of the terminal to ensure its SIM is registered on the network and the terminal is successfully transmitting data. See the preceding figure for an example of this interface.

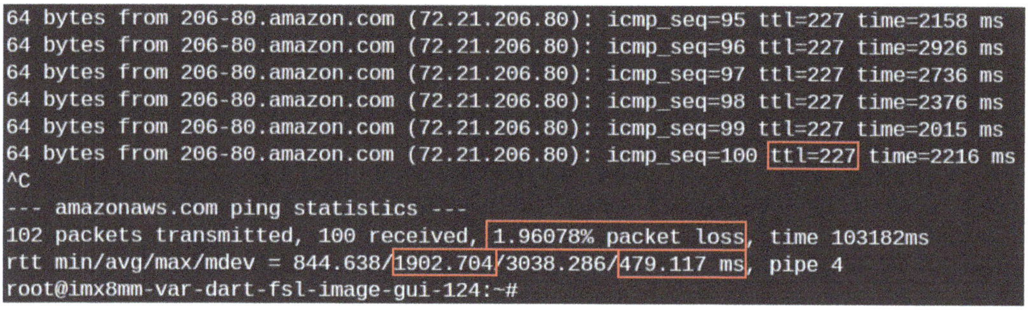

```
64 bytes from 206-80.amazon.com (72.21.206.80): icmp_seq=95 ttl=227 time=2158 ms
64 bytes from 206-80.amazon.com (72.21.206.80): icmp_seq=96 ttl=227 time=2926 ms
64 bytes from 206-80.amazon.com (72.21.206.80): icmp_seq=97 ttl=227 time=2736 ms
64 bytes from 206-80.amazon.com (72.21.206.80): icmp_seq=98 ttl=227 time=2376 ms
64 bytes from 206-80.amazon.com (72.21.206.80): icmp_seq=99 ttl=227 time=2015 ms
64 bytes from 206-80.amazon.com (72.21.206.80): icmp_seq=100 ttl=227 time=2216 ms
^C
--- amazonaws.com ping statistics ---
102 packets transmitted, 100 received, 1.96078% packet loss, time 103182ms
rtt min/avg/max/mdev = 844.638/1902.704/3038.286/479.117 ms, pipe 4
root@imx8mm-var-dart-fsl-image-gui-124:~#
```

Figure 9.4 – Connection quality from the terminal to AWS

Once we have a good connection, we can do some `ping` tests to `amazonaws.com` from our laptop. Looking at the preceding figure, you'll notice a few things. First, the average RTT is 1,902 ms – nearly 2 whole seconds. This is long enough to cause outright timeouts for some applications, and very low throughput for anything TCP-based. Second, the mean deviation of 479, or almost 25%, represents an extreme amount of jitter. Finally, while a packet loss of nearly 2% would be considered extremely high and unacceptable over a terrestrial internet connection, it is not uncommon over geostationary SATCOM links such as this.

These are important considerations for services such as DNS and NTP – you should consider running local versions of these services and tuning the caching behavior to prevent the need to reach over the SATCOM link for every single address resolution.

Now, we can power up, unlock, and configure AWS Snowcone using AWS OpsHub or the CLI per the `AWS Snow family documentation`. Note that you must deploy the preinstalled Greengrass AMI on an `snc1.medium` instance.

Understanding AWS IoT Greengrass relationships

The following figure provides an overview of the relationship between the multiple logical constructs you must configure within the AWS IoT Core and AWS IoT Greengrass services in-region. Remember, all the Greengrass v2 agent does is dial home and fetch its configuration from AWS. This includes the components (software) it needs to download from S3 and instructions on how to set them up:

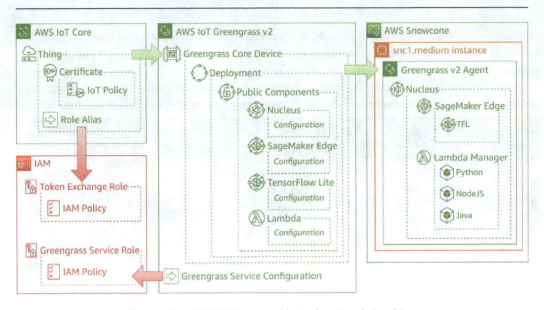

Figure 9.5 – AWS IoT Greengrass logical entity relationships

AWS IoT Core things and AWS IoT Greengrass Core devices

In AWS IoT Core, all devices are things. In this architecture, AWS Snowcone is a Greengrass Core device, but first and foremost, it is a thing in IoT Core (albeit of a special type). In the following screenshot, we have three things configured – two of them are sensors that report to the Greengrass v2 agent on our Snowcone, and one of them is the Snowcone itself:

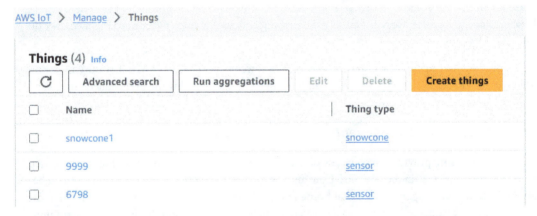

Figure 9.6 – AWS Snowcone and sensors as things in AWS IoT Core

The following screenshot shows an overview of snowcone1 from the perspective of AWS IoT Greengrass as a Greengrass Core device:

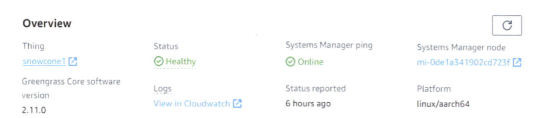

Figure 9.7 – Overview of snowcone1 as an AWS IoT Greengrass Core device

AWS IoT Greengrass deployments and components

You will notice that we have configured this particular device to be an AWS Systems Manager node, as well as set it up to push detailed logs and metrics to AWS CloudWatch. We did this by targeting it with a Greengrass deployment that contains public Greengrass components. Examining the following screenshot, you will see that we've also deployed several other Greengrass components. These include the things we need to run AWS Lambda functions, a local Moquette-based MQTT broker, and the MQTT bridge, which relays messages from the local broker up to matching topics in AWS IoT Core:

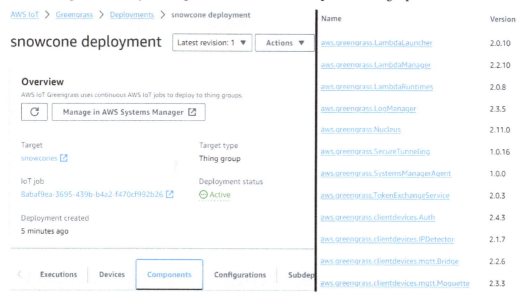

Figure 9.8 – A Greengrass deployment and its constituent Greengrass components

AWS IoT certificates and policies

All things – regardless of type – are mapped to individual certificates in AWS IoT Core. This goes for our AWS Snowcone device, as well as the IoT devices reporting to it. These certificates can be handed out by a private CA (either one of your own, or one in AWS Certificate Manager), but more commonly, they are provisioned by the internal CA the AWS IoT Core service itself uses. That is what we are doing in the following screenshot:

Figure 9.9 – An AWS IoT certificate and the IoT policy attached to it

You will also notice that the AWS IoT certificate for snowcone1 has an IoT policy attached to it. This is a type of policy distinct from the ones used in AWS IAM. The following example JSON shows a sample IoT policy of this type. You will notice that it governs only what the thing can do in the context of IoT and Greengrass actions:

```
{
  "Version": "2012-10-17",
  "Statement": [
    {
      "Effect": "Allow",
      "Action": [
        "iot:Connect",
        "iot:Publish",
        "iot:Subscribe",
        "iot:Receive",
        "iot:Connect"
      ],
      "Resource": "arn:aws:iot:eu-west-2:XXXXXXXXXXXX:*"
    },
    {
      "Effect": "Allow",
      "Action": "greengrass:*",
      "Resource": "*"
    },
    {
      "Effect": "Allow",
      "Action": "iot:AssumeRoleWithCertificate",
      "Resource": "arn:aws:iot:eu-west-2:XXXXXXXXXXXX:rolealias/
snowcone-role-alias"
    }
  ]
}
```

Figure 9.10 – Example of an IoT policy

Notice the last section of the preceding policy. The resource it points to is another AWS IoT construct called a role alias. This is simply a mapping of the role alias ARN to the ARN of a standard IAM role. This is how we allow the Greengrass Core device's certificate to assume a standard AWS IAM role for non-IoT/Greengrass-specific operations. This might include things such as permissions to publish to AWS CloudWatch, access certain Amazon S3 buckets, perform actions in AWS Systems Manager, or perform an `iam:PassRole` action to take on the persona of yet another IAM role.

Passing MQTT messages through the IoT gateway

Now that we've given you a general overview of how AWS IoT Greengrass itself is put together, we will walk through options for connecting IoT devices such as sensors and cameras as things behind the Greengrass Core device.

Local MQTT broker and the MQTT bridge

IoT devices can send MQTT messages directly to AWS IoT Core in the cloud via the public endpoints in the nearest AWS region. However, if they prefer to relay messages via a Greengrass Core device, they need to target an MQTT broker running locally on it. Two public components can be pushed out to a Greengrass Core device for this purpose. One is based on Moquette (for MQTT v3.1.1) and the other is EMQX (for MQTT v5). Regardless of which the customer chooses, by default, when a device sends MQTT messages to the local broker, they will just sit there doing nothing unless another component grabs them and acts:

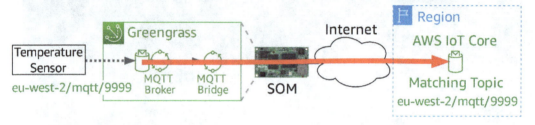

Figure 9.11 – An MQTT broker relaying messages to AWS IoT Core via the MQTT bridge component

The preceding figure gives an overview of how the MQTT bridge can be inserted to relay these messages to matching topics in the AWS region. In the Greengrass deployment, the MQTT bridge component is added and configured to listen to specific topics on the local MQTT broker. It will automatically pick them up and insert them into matching topics in AWS IoT Core:

```
{
  "reset": [],
  "merge": {
    "mqttTopicMapping": {
      "iot-core-mapping-eu-west-2": {
        "topic": "eu-west-2/+/+",
        "source": "LocalMqtt",
        "target": "IotCore"
      }
    }
  }
}
```

Figure 9.12 – MQTT bridge component configuration example

Take a look at the preceding configuration example. The `"topic"` parameter is set to listen to whatever local MQTT broker is running for topics matching a certain pattern. The `"target"` parameter tells it where to replicate them to.

This is a simple example of how an AWS IoT Greengrass Core device can aggregate MQTT messages, regardless of the connectivity status back to the AWS region. Greengrass will keep trying to talk to whatever endpoint in AWS it is configured for until the path comes up, at which time any messages the bridge has picked up will be pushed to AWS IoT Core. If the uplink is down for a week, the bridge will keep buffering messages in this way – so long as it has sufficient local resources to store them.

Embedded Linux devices as IoT gateways

Millions of smart devices around the world are built on AWS IoT Greengrass. From doorbell cameras to washing machines to industrial HVAC units, Greengrass is deployed at a massive scale – the quiet hero of IoT.

Let's elaborate a bit on two types of embedded device solutions based on AWS IoT Greengrass:

- **Wearable devices for first responders**: First responders, relief workers, firefighters, and the like are often deployed en masse to dangerous environments. In many cases, the workers themselves are dispersed and not aware of the larger picture. Wearable sensors, cameras, and panic buttons reduce the time needed to assist a worker in distress, as well as preserve evidence and direct resources to where they can do the most good. Running inexpensive single-board computers dramatically reduces the time to market for such devices.

  The base station they report into can be rapidly developed using `AWS IoT Core for LoRaWAN` – an AWS-managed service based on `Semtech's LoRa Basics Station` and network servers that run inside AWS IoT Core. Customers can build gateways using commodity hardware (see this `tutorial for doing so on a Raspberry Pi`). In addition, the AWS Partner Device Catalog `has dozens of listings for prebuilt and certified hardware` based on this service.

- **Home monitors for energy efficiency**: Energy providers want to instrument customer's energy usage patterns and educate them on ways they could lower their consumption. To do this, an inexpensive single-board computer hosting the AWS IoT Greengrass agent is attached to the breaker panel in the customer's home. It has small external clamps on each circuit, and the circuit names (kitchen, laundry, living room, and so on), along with appliance model information, are posted to the central management portal by the technician doing the installation. Current draw readings are aggregated with any incoming streams from IoT-capable appliances and sent back in daily batches over an inexpensive cellular NB-IoT connection. All of this data is used to train an ML model in AWS SageMaker in the cloud, and over time, the model develops an ability to make recommendations directly to users in their homes via Alexa-capable speakers or the Alexa mobile application. For instance, it could alert the user that they have left the garage light on past 9 P.M., something that normally does not occur. Or, based on the model's understanding of a vast array of appliance models, it could suggest a specific model of refrigerator based on how much money it will save the customer on their bill versus the cost of replacement.

The Yocto Project

You are likely aware that many of the smart devices in your home or workplace run on Linux. If you've ever wondered what distribution they run, the answer is probably none. They probably use a custom-compiled Linux kernel that minimizes its footprint and maximizes its security profile. This is typically accomplished via the Yocto Project.

The Yocto Project is an open source initiative that facilitates the creation of custom Linux distributions tailored for embedded systems and IoT devices. Rather than being a Linux distribution itself, it provides a framework and tools that allow developers to create their own distributions, optimized for specific hardware configurations or applications.

At the heart of the Yocto Project is the Poky build system, which uses the BitBake tool, a task executor and scheduler that is used to manage and create custom Linux distributions. It provides a layer-based architecture, which means that various components, such as the kernel, user-space applications, and metadata, are kept separate, making it easier to modify and manage individual aspects of the distribution.

Developers can use Yocto to create a lightweight, streamlined Linux distribution with only the necessary components for their specific use case, thereby reducing the footprint and improving the performance of the resulting system. This customization makes it a preferred choice for embedded systems where resources are limited and optimization is key.

AWS is a Platinum member of the Yocto Project. This means you know the `meta-aws` layer (available `directly in Yocto's repository`) is extensively tested, stable, and supported with any new Yocto Project releases.

Fleet Provisioning

The AWS IoT Greengrass Core v2 agent can be installed into Linux and registered by hand into a specific AWS account every time:

```
useradd --system --create-home ggc_user
groupadd --system ggc_group
echo "ggc_user ALL=(ALL:ALL) ALL" >> /etc/sudoers


curl -s https://d2s8p88vqu9w66.cloudfront.net/releases/greengrass-
nucleus-latest.zip


unzip ./greengrass-nucleus-latest.zip -d ./GreengrassInstaller

java -Droot="/greengrass/v2" -Dlog.store=FILE && \
-jar ./GreengrassInstaller/lib/Greengrass.jar && \
--aws-region eu-west-2 && \
--thing-name device-88574 && \
--thing-group-name thing-group-eu-west-2 && \
--component-default-user ggc_user:ggc_group && \
--provision true && \
--setup-system-service true && \
--deploy-dev-tools true
```

Figure 9.13 – Manual installation of the Greengrass v2 agent into Linux

The preceding code shows the procedure for manual installation. Keep in mind that before you run this, you must install the AWS CLI and configure it with credentials that have permission to perform all the steps needed to create the necessary objects in the target AWS account. For a large fleet of devices, this is labor-intensive and arguably insecure.

The Fleet Provisioning feature is an alternative approach that generates a set of claim certificates that are the same for all devices in a fleet – at first, anyway. Once the device boots for the first time, it reaches out to the AWS IoT Greengrass endpoints it has been configured for and exchanges its claim certificate for an individual certificate that uniquely identifies that device. From there, it operates in the same way as a manually provisioned device would. This is particularly useful for embedded Linux builds that use Yocto, which is why it is included in the meta-aws layer.

The following code shows an example of the configuration needed to install the Greengrass v2 agent and set up Fleet Provisioning for the local.conf file when doing a Yocto build that includes the meta-aws layer:

```
IMAGE_INSTALL:append = "greengrass-bin"
PACKAGECONFIG:pn-greengrass-bin = "fleetprovisioning"
GGV2_DATA_EP = "a31sample-ats.iot.eu-west-2.amazonaws.com"
GGV2_CRED_EP = "c21sa.credentials.iot.eu-west-2.amazonaws.com"
GGV2_REGION = "eu-west-2"
GGV2_THING_NAME = "device-88574"
GGV2_TES_RALIAS = "greengrass-token-exchange-role-alias"
GGV2_THING_GROUP = "thing-group-eu-west-2"
```

Figure 9.14 – Example of Fleet Provisioning configuration in local.conf for bitbake/Yocto

Moving Greengrass to the SATCOM terminal

Imagine a situation like we had in the previous section where we want to put an IoT gateway in the middle of nowhere behind a SATCOM terminal, but for various reasons, AWS Snowcone isn't ideal. Maybe we intend to deploy it once and never physically visit the site for 5 years. How could we run the AWS IoT Greengrass v2 agent on the terminal itself?

As you'll recall from earlier in this chapter, most SATCOM terminals have just enough local compute resources to run a **Real-Time Operating System** (**RTOS**) that hosts a web interface for configuration and orchestrates digital signal processing tasks. There is a FreeRTOS version of the Greengrass v2 agent. We *might* be able to squeeze it onto the existing microprocessor onboard the terminal. But for an IoT gateway that will support many different types of sensors, and especially one that will run an ML model, we need more horsepower.

We could augment the terminal with a single-board computer such as a `Raspberry Pi Zero 2 W`, or one of the `more powerful modules available from companies such as Variscite`. We could simply install Linux normally on top of it, but to conserve resources and simplify mass deployment, we would prefer to create a custom build of embedded Linux containing the AWS IoT Greengrass agent, Fleet Provisioning certificates, and bootstrap configuration scripts. This would allow us to flash microSD cards on a per-deployment basis, inserting Fleet Provisioning certificates and endpoint configuration specific to a group of devices we want attached to a certain AWS account. It would also allow the user to receive the terminal, point it, and power it on – the rest of the configuration is handled by AWS IoT Greengrass components that configure the terminal environment:

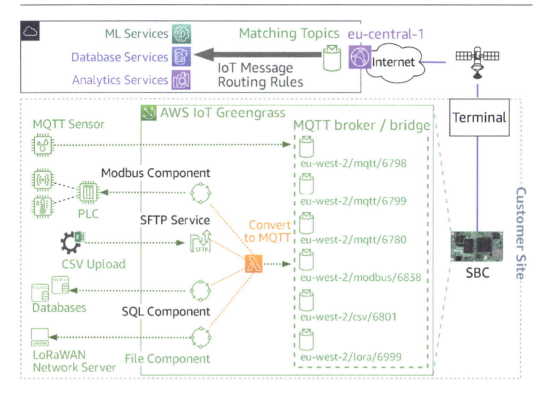

Figure 9.15 – SATCOM terminal enhanced with edge compute using AWS IoT Greengrass

Next, the Greengrass v2 agent communicates with an endpoint in AWS and exchanges its Fleet Provisioning certificates for a true individual certificate with an IoT policy and IAM roles attached to the terminal vendor's SAAS platform account in AWS. At that point, whatever edge compute configuration the customer has subscribed to is pushed down via components of a Greengrass deployment. Now, the customer can directly point IoT sensors at a local MQTT queue on the terminal, upload CSV data files to it via SFTP, set up poll/response configurations for Modbus PLCs, and so on.

Tactical edge

Tactical edge refers to edge computing environments that support military operations. Such situations are characterized by limited connectivity, high mobility, and the most intense security requirements.

At the tactical edge, the most common needs include deploying, managing, and securing the following technologies:

- **IoT**: Cameras and sensors of every description get cheaper every day. The number of data-gathering devices is growing just as quickly on the battlefield as it is back home. These devices need to either facilitate immediate/local decision-making or move the data they collect to an aggregation point where those things can be done.

  Compared to prior decades, when such things would have been done using proprietary protocols and architectures specific to each system, nowadays, the same IoT protocols used in the civilian world are used at the tactical edge.

  For example, consider a UAV that drops thousands of LoRaWAN-capable vibration sensors over several square miles of terrain around a **Forward-Operating Base** (**FOB**). These could leverage LoRaWAN's inbuilt geolocation features to triangulate where a threat is coming from without each sensor requiring GPS units of their own. A system like this is so cheap that the unit probably wouldn't even bother to collect the sensors when they moved on.

- **ML**: One of the most vital roles of a communications officer is to facilitate the rapid movement of information from where the action is to where the intelligence teams can analyze it. Those teams have an ever-increasing amount of data coming at them in the form of audio, video, and IoT sensor data. How can they make sense of it all? How can they know whether a given individual walking by a camera is a known threat or someone to be protected? How do they quickly translate and transcribe thousands of conversations captured by microphones attached to those cameras? Military intelligence teams are relying more and more on ML to perform such tasks, freeing up analysts and enabling them to better advise commanders.

  ML models are also being used for the same sorts of things seen in the civilian world. This includes V2X-like monitoring and control of vehicles, condition monitoring of equipment for failure prediction, or optimization of supply chains.

- **Containers and virtualization**: Much like the situation with IoT, the applications needed by military commanders are growing exponentially – and militaries around the world are responding by standardizing on the same tools used in the civilian world to do DevOps, DevSecOps, MLOps, and so on.

Forward deployment of the AWS Snow family

Commanders need flexible, rugged compute horsepower that can be quickly deployed or torn down and moved. From an off-the-shelf server's point of view, forward-deployed **Command Posts** (**CPs**) operate under less-than-ideal circumstances. Sometimes, they are semi-permanent structures that have HVAC, but even then, it is limited. CPs rarely meet the environmental requirements of the average rackmount or tower server. Snowball Edge devices meet stringent ruggedization standards, including MIL-STD-810G. This, combined with the ability to use the same APIs and constructs their applications use on AWS in-region, makes them attractive in such situations.

To meet the need for CPU/GPU resources in the CP, customers often deploy multiple Snowball Edge devices. However, Snowball Edge devices themselves do not have a notion of clustering. They do not form a relationship with each other at the infrastructure level the way hyperconverged solutions from Nutanix or Cisco would. While they can freely communicate with each other over the local network, each node is an island of capacity unto itself.

In such circumstances, **Amazon Elastic Kubernetes Service Anywhere** (**EKS-A**) can be overlaid on top of multiple AWS Snowball Edge devices to provide these things to a certain degree.

Very similar needs are observed in disaster response scenarios. `AWS routinely uses AWS Snow family devices to provide disaster relief` to affected communities around the world.

EKS-A

EKS-A is an extension of Amazon's highly popular managed Kubernetes service that facilitates the deployment, management, and scaling of containerized applications. EKS-A allows customers to run Kubernetes clusters not just in AWS regions but also in their own data centers, other cloud environments, or even in DDIL scenarios.

AWS EKS Distribution

AWS EKS Distribution and EKS-A are both elements of AWS's broad strategy to provide flexible and robust Kubernetes solutions, but they serve different roles in the ecosystem.

AWS EKS Distribution is the same Kubernetes distribution that is used by all variants of EKS (both in-region and EKS-A). It is `freely available on GitHub` for anyone to deploy in their environment. The distribution provides a consistent set of binaries vetted by AWS, allowing anyone to set up and manage a Kubernetes environment on their own – with the reassurance that it adheres to AWS's standards of security and reliability. It does not include integrated tooling for cluster creation and management.

On the other hand, EKS-A is a more comprehensive solution that is built upon the AWS EKS Distribution. It not only includes the core Kubernetes distribution but also provides a fully integrated and automated tooling system for deploying and managing Kubernetes clusters in any environment, be it on-premises or in other clouds.

EKS-A includes things such as an installer and command-line tool that help users create, upgrade, and manage Kubernetes clusters. It also offers an optional connector that can talk back to the EKS service endpoints in-region, allowing customers to visualize the status of remotely deployed EKS-A clusters.

EKS-A container networking interfaces (CNIs)

Because disconnected environments do not participate in the network virtualization stack backing in-region VPC services, the VPC-CNI plugin is not available. Instead, EKS-A offers a choice of Cilium or Kindnet for container networking. Only one can be selected at deployment time, and it cannot be changed after deployment:

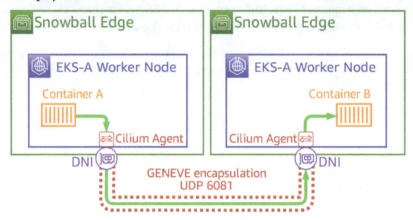

Figure 9.16 – EKS-A encapsulating IP flows between worker nodes using GENEVE

EKS-A providers

Similar to how EKS handles the deployment/management of EC2 instance-based worker nodes in-region, EKS-A supports several providers that allow it to do something similar with non-EC2 compute resources.

The supported providers include the following:

- **VMware vSphere**: Via a vCenter central management server
- **Nutanix AHV**: Via a Prism central management server
- **CloudStack**: Via a cluster endpoint (similar to the K8S API)
- **Bare metal**: Via BMC/IPMI interfaces to orchestrate network builds
- **Snow**: Via AWS Snow APIs on each Snow family device
- **Docker**: For unsupported development environments only

This section will cover scenarios involving the bare metal and Snow family providers only. Detailed information about all providers can be found in the `installation section of the EKS-A documentation`.

The EKS-A provider for Snow family devices

The EKS-A provider for Snow family devices allows customers to deploy and manage EKS-A clusters on AWS Snowball Edge devices. An entire EKS-A cluster can be deployed to a single device for maximum portability and simplicity, or it can target three or more devices to provide higher availability and scaling of compute resources:

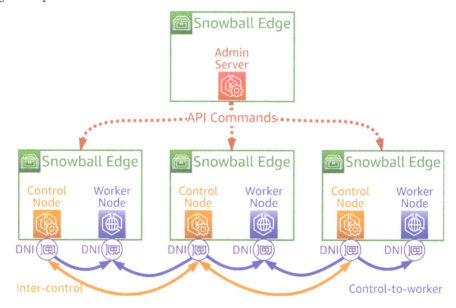

Figure 9.17 – Amazon EKS-A Snow family provider communication channels

Regardless of the chosen topology, the deployment process is essentially the same for a fully disconnected environment:

1. **Order AWS Snowball Edge device(s)**: When selecting the type of device, choose one of the three Snowball Edge compute-optimized variants during *Step 2*. After doing so, the *Step 3* page will show a tab that lets you choose options for EKS-A on Snow. This includes whether you wish to have everything on a single device or a cluster of up to 10 devices. This is also where you can select the AMI and Kubernetes versions to install on the node images.

   When the devices arrive, connect to and unlock them with AWS OpsHub or the Snow CLI as normal and follow the manual setup procedure found in the `AWS Snow Family documentation`. Alternatively, setup tools are offered that automate the setup process. See the `setup-tools section of the EKS-A documentation` for more details.

2. **Set up a local harbor registry instance**: This is needed in DDIL scenarios to eliminate the need for EKS-A to reach back to the AWS **Elastic Container Registry** (**ECR**) to obtain images. See the `container-registry-ami-builder` section of the EKS-A documentation for more information.

3. **Instantiate the EKS-A admin server and deploy EKS-A clusters**: The admin server is a small standalone EKS cluster that is used to bootstrap the environment and deploy/manage other clusters that contain your workloads. See the `Install on Snow` section of the EKS-A documentation for more specifics.

Direct Network Interfaces (DNIs) and EKS-A

When EKS-A deploys cluster nodes to AWS Snowball Edge devices, it creates and associates a DNI to each node. This allows pods within the nodes to use GENEVE encapsulation with the Cilium CNI – thus avoiding the need for NAT. While this is desirable, it limits the number of nodes per device due to the constraint of seven DNIs per physical interface (see *Chapter 4*).

EKS – a provider for bare-metal servers

Some disconnected environments require the use of a specific server vendor's hardware for one reason or another. In such cases, the EKS-A provider for bare-metal servers can be used similarly to the provider for the AWS Snow family:

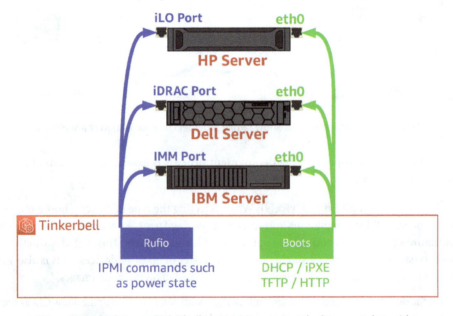

Figure 9.18 – EKS-A uses Tinkerbell to provision against the bare-metal provider

The key difference between the two providers is the absence of the AWS-managed APIs for out-of-band management that the AWS Snow family offers. Instead, it leverages the `open source provisioning system created by Equinix called Tinkerbell`. As shown in the preceding figure, Tinkerbell consists of multiple services that do things such as netboot/kickstart-style provisioning or control the power state of the server via whatever out-of-band management facilities that the server vendor offers.

See the `Install on Bare Metal section of the EKS-A documentation` for more information.

Forward deployment of AWS Outposts

As you'll recall from our discussion of AWS Outposts, the site requirements are typical of a corporate data center or colocation facility. These are often difficult to accommodate in remote areas. For instance, the temperature and humidity must be kept within a tight range – not a simple proposition when a densely populated rack of servers generates as much as 50,000 BTUs an hour. This requires heavy-duty HVAC to maintain. Similar problems emerge when considering high amounts of well-conditioned AC power:

Figure 9.19 – AWS Modular Data Center

Enter AWS **Modular Data Center** (**MDC**). MDC is a self-contained mini-data center that meets MIL-STD-810H requirements and can be deployed to remote locations. Each MDC unit consists of two shipping containers – one of which contains the equipment needed to maintain the environment (including redundant HVAC and power), while the other can accommodate up to five AWS Outposts racks.

Perhaps most importantly, all of the environmental subsystems involved can be remotely monitored and managed by the customer. For example, an alert would go out if one of the HVAC units failed and one of the spares had to be brought online.

At the time of writing, MDC leverages MEO/GEO SATCOM provider SES' `Cloud Direct` `offering` to deliver consistent connectivity back to the AWS parent region needed for the AWS Outposts service link.

Private 5G and DDIL

`AWS Private 5G` is a managed service that takes advantage of the open CBRS spectrum, which is only available in the US. It is similar in architecture to AWS Outposts – indeed, it can involve AWS Outposts rack in some cases. This means significant components of the service run in an AWS region, thus necessitating clean and consistent high-speed connectivity back to a region in the US. Due to this, it is not suitable for DDIL situations.

Many DDIL scenarios can benefit from an isolated 5G network, even in the absence of good enough connectivity to support AWS Private 5G – or any backhaul at all. Solutions for these situations can be built with AWS services such as the AWS Snow family combined with a partner product that runs on top such as those available from Athonet or Federated Wireless.

As we discussed in *Chapter 2*, network function virtualization based on containerization is the de facto standard upon which 5G core software is built. Thus, we need more than just the AWS Snow family for these situations.

Using AWS Snowball Edge to host a private 5G core

The following figure demonstrates a typical architecture for a mobile 5G core. A containerized NFV partner solution is installed into EKS-A, which, in turn, is deployed onto a single AWS Snowball Edge device. The RAN provided by the partner ISV in this particular case includes the RAN functions plus the RF equipment in a single piece of hardware:

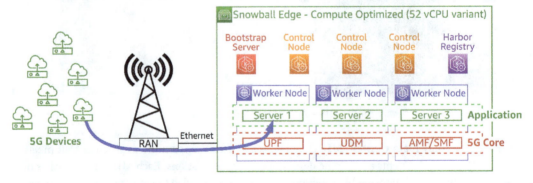

Figure 9.20 – Mobile 5G core on AWS Snowball Edge

This architecture is not uncommon in such circumstances as it simplifies the management of the 5G core in the sense that you always know that all components are on this one device. However, this lack of redundancy sacrifices resilience at the physical layer. It also services a limited number of 5G devices.

You could expect this smaller variant of the compute-optimized AWS Snowball Edge to support the base EKS-A components, a small 5G core, and a small user-built application. In turn, this small 5G core would probably support ~100 5G devices registered on the RAN. However, the aggregate throughput between them and the application servers will be constrained by the Ethernet port(s) used on the AWS Snowball Edge device, and the applications they are communicating with. In other words, you would probably not expect all 100 devices to be able to push 1 Gbps to the application simultaneously, even if the devices individually could do so.

Summary

In this chapter, we explored various scenarios involving DDIL settings. Initially, we presented an overview of these conditions, describing the challenges that are encountered under DDIL conditions.

Next, we addressed the utilization of AWS IoT services in DDIL contexts. We examined how AWS Snowcone operates as an efficient IoT gateway that facilitates data processing and storage in edge locations and reaches back to AWS over the Inmarsat BGAN SATCOM network. Moreover, we gave an overview of AWS IoT Greengrass, illustrating how it integrates with other AWS services to provide a cohesive solution in limited connectivity areas.

Following that, we moved on to the tactical edge segment, focusing on the deployment strategies involving the AWS Snow family and AWS Outposts. We described the benefits of integrating EKS-A with both AWS Snowball Edge and bare-metal servers, providing a resilient and rugged solution. Then, we discussed AWS MDC and how it hosts AWS Outposts in forward-deployed locations, enabling seamless operations in such environments.

Lastly, we turned our attention to the intersection of private 5G networks and DDIL scenarios. Here, we highlighted how EKS-A and AWS Snowball Edge combine with partner ISV solutions to build a private 5G core. This allows customers to quickly deploy isolated 5G networks to support communication with devices under circumstances where Wi-Fi will not suffice.

In the next chapter, we will take a look at how customers are combining AWS Wavelength with other AWS services to build **Multi-Access Edge Computing** (**MEC**) solutions.

10

Utilizing Public 5G Networks for Multi-Access Edge (MEC) Architectures

As you'll recall from *Chapter 2*, **Multi-Access Edge Computing** (**MEC**) refers to architectures that exploit the opportunity presented by the convergence of two forces in 5G architecture. First is the prevalence of **Network Functions Virtualization** (**NFV**) on commodity server hardware. Second is the fact that the **User Plane Function** (**UPF**) is distributed at the edge of 5G networks – as opposed to the centralized **Packet Gateway** (**PGW**).

MEC use cases we'll review include online gaming, **Computer Vision** (**CV**), robotics, **vehicle-to-everything** (**V2X**), and new techniques for video production at live events. For publicly accessible systems running on **Mobile Network Operator** (**MNO**) networks, we'll focus on AWS Wavelength. For private MEC solutions, we will cover using AWS Outposts and AWS Private 5G in the US, and the Integrated Private Wireless program in other regions.

In this chapter, we're going to cover the following main topics:

- Overview of architecting 5G-based MEC solutions
- Observability, security, and capacity of Wi-Fi versus 5G
- Computer Vision
- V2X
- Software-defined video production

Overview of architecting 5G-based MEC solutions

Imagine you are an architect tasked with designing a city-wide public safety solution for a major metropolitan area. You need to install **Point-Tilt-Zoom** (**PTZ**) cameras in strategic locations around the city. They must be positioned in such a way as to be able to target and track individual vehicles, pedestrians, or even drones in response to inferences from an ML model running in the cloud. Each camera module is capable of streaming multiple types of high-definition (8K) video simultaneously over `RTSP` – including optical, infrared, and `LiDAR`. Separately, they each have a **command-and-control** (**C2**) channel that must always function:

Figure 10.1 – Ruggedized PTZ multispectral 5G-capable camera system

As you walk around the city, identifying the best places to mount these systems, you realize getting high-speed internet access to them is going to be a challenge. One might go on the same pole as a street lamp. The next might need to be mounted on the side of a building three stories up. Another may need to be placed on its own pole next to a construction site.

You begin figuring out how to accomplish this by contacting multiple local broadband providers and determining which ones can cover which camera location. Each provider offers different speeds, different SLAs, different costs, and different technologies. Some will run cables all the way to your camera. Others take one look at where you want the device and say, "*sorry, we aren't licensed to do this.*" You realize this is going to take a long time, be expensive, complex to manage, and some locations simply won't meet the SLAs needed for either the control channel uptime requirement of 99.99%, while others won't be able to ensure the 200 Mbps required for all possible RTSP streams. Then, the project manager pokes their head in your office and says, "*by the way, we need to add 50 mobile cameras that can be set up and torn down by the police within 30 minutes each.*" This is a problem.

A scenario such as this is perfect for 5G. Recall from *Chapter 2* the discussion about network slicing. A 5G provider could provision each camera such that it has a network slice for HTTPS traffic that is limited to 5 Mbps but has an availability SLA of 99.99%, while a second network slice supports 300 Mbps but has a lower availability SLA of 99%. All cameras would use the same technology and get the same SLA. Better still, they can be powered up and connected immediately:

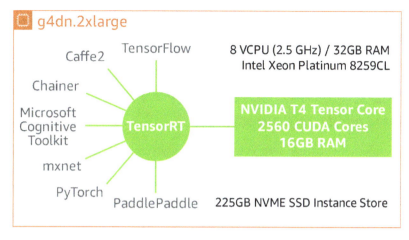

Figure 10.2 – GPU-accelerated EC2 instances available in AWS Wavelength

Combine 5G network slicing with the g4dn.2xlarge EC2 instances available in that metropolitan area's AWS Wavelength Zone, and you've got an MEC-based solution that puts the ML model very close – at latencies as low as 1 ms (depending upon the MNO and your network slicing config).

Public MEC

The following diagram illustrates an AWS Wavelength Zone and customers in the surrounding metropolitan area using it. AWS Wavelength is a public MEC solution, meaning any customer off the street can buy a SIM card from Vodafone, Verizon, KDDI, Bell, SK, or BT as appropriate and start using an EC2 instance or ECS/EKS container in the associated AWS Wavelength Zone:

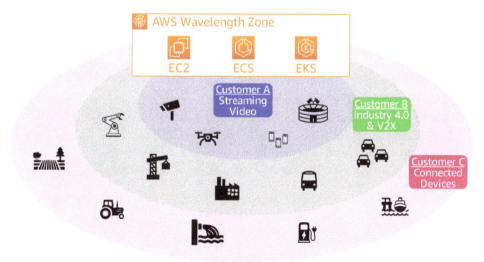

Figure 10.3 – MEC with AWS Wavelength

Customers use 5G to build systems for many different purposes within a given metropolitan area, while the compute infrastructure at the 5G edge is shared among them. The underlying resources (AWS Outposts and AWS Direct Connect) supporting the EC2 instances or ECS/EKS containers are managed by the MNO, which offers the 5G service itself, thus enabling optimizations such as network slicing that extend from the customer devices to the compute resources involved.

Multicarrier and non-5G interoperability

Everything we've covered so far has been in the context of public MEC on one carrier – but there's no reason an application has to be limited to a single 5G network. Nor is there a law saying all users of a given application must connect to it over 5G at all. You could architect a globally distributed application that allows certain users to benefit from the high speed and low latency available to 5G devices in appropriate regions. At the same time, other users connect via whatever means they have available, accepting higher latency:

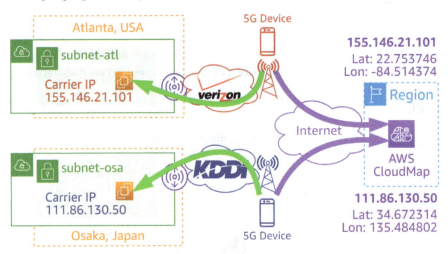

Figure 10.4 – Multi-carrier public MEC architecture with AWS Wavelength

The preceding diagram shows a simple example of this where we assume the application on user devices has permission to query the user's location. The client's first step is to ask AWS Cloud Map which carrier IP they should attempt to reach given their latitude and longitude. This will yield response times as low as a single millisecond – the most ideal case:

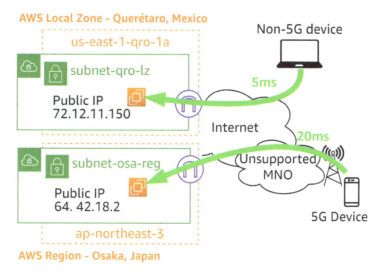

Figure 10.5 – Fallback paths for users not on supported MNO networks

The preceding diagram shows what happens for other users of the application. Clients on a 5G device with a GPS position that matches an AWS Wavelength Zone would first try the carrier IP handed to them by AWS Cloud Map. If they can't reach it, that must mean they aren't on the correct MNO's network. In this case, the client would then fall back to the next best thing, which in this case is a public IP behind an internet gateway in a standard AWS Region. Clients who are not on 5G at all will route through whatever internet connection they have to the closest resource; in this case, an AWS Local Zone.

This architecture would result in a wide range of user experiences depending on the location and client technology. AWS has been helping architects of multiplayer online games address these challenges for years. Such disparities in player Round Trip Time (RTT) are only increasing with the introduction of 5G-capable gaming devices such as the Razer Edge 5G sold by Verizon:

Figure 10.6 – Razer Edge 5G gaming device sold by Verizon

AWS has solutions that allow game builders to accommodate this type of device into a globally accessible game in the cloud. Game designers can create a situation where users of devices such as this that have 1 ms RTT to the game server are not overly advantaged compared to users that have a 30 ms RTT via their home internet connection.

You can think of Amazon GameLift as an **Auto-Scaling group** (**ASG**) that is specifically designed for the protocols involved with multiplayer games that span multiple geographies. It will launch additional game servers where needed based on demand. It also has a feature called FlexMatch that will group players with similar characteristics (such as RTT/latency) to ensure a level playing field.

The following diagram shows an example of an architecture for an online game that integrates these concepts:

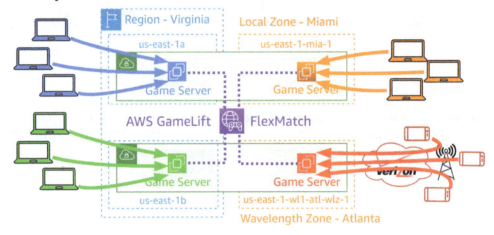

Figure 10.7 – Using Amazon GameLift FlexMatch to send players to the best game servers

Private MEC

Some customers wish to build their own private 5G networks and offer MEC to their customers only. Often, these customers are subordinate companies of an overarching conglomerate, or sometimes they are public sector entities. For instance, the State of California might decide to offer MEC services only to county- or municipal-level entities within certain geographies:

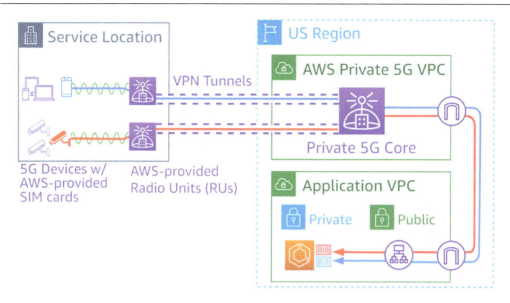

Figure 10.8 – Devices accessing ECS containers over AWS Private 5G

The preceding diagram illustrates a way to implement private MEC using the AWS Private 5G service with customer applications hosted on ECS containers in-region. As with any other MEC solution, we can take advantage of the fact that additional general-purpose compute is available in the same place the 5G UPF is hosted via NFV.

Observability, security, and capacity of Wi-Fi versus 5G

When considering general-purpose wireless technologies for a new architecture, the decision is typically 802.11x-based Wi-Fi or 5G. When both are available, which one makes the most sense is usually driven by the observability, security, and capacity needs of the application in question.

Observability

Wi-Fi networks, typically deployed in confined areas such as homes or offices, face challenges related to interference, especially in densely populated environments. Observability can be hindered due to overlapping channels, signal interference from physical obstacles, and the presence of multiple devices competing for access. Tools for Wi-Fi observability, such as network analyzers, provide insights into network health but are expensive, and the onus often lies with local administrators to monitor and maintain optimal performance.

These tools were developed in response to needs in the field, in most cases long after the 802.11x specifications were developed. As an architect, when you implement a Wi-Fi-based solution, you must integrate these tools and build associated reporting mechanisms for them from the start – they must not be an afterthought or your user experience will almost certainly be poor, and intermittently poor at that.

5G, on the other hand, was developed by a consortium of MNOs. These are **Service Providers (SPs)** who maximize profits by centralizing such things as much as possible. Therefore, sophisticated QoS mechanisms such as network slicing are heavily instrumented, and the expensive tooling is centralized and exposed to customers via standard APIs.

What this means for you as an architect is you can simply tell the MNO you are working with what network slices and performance SLAs you need for each. They provision them, and you are given an API-based reporting interface that proves each slice on each device is both available and performing to the level you are paying for.

Security

Security has been a perennial concern for Wi-Fi. Early encryption methods such as **Wired Equivalent Privacy (WEP)** were quickly found to be vulnerable. While the subsequent WPA and WPA2 protocols offered enhanced security, they were not impervious to breaches. The introduction of WPA3 has aimed to address earlier vulnerabilities. However, Wi-Fi often remains susceptible to threats such as **Man-In-The-Middle (MITM)** attacks, especially in public networks. Users must rely on network administrators to regularly update and patch router firmware to fend off potential threats.

5G was designed with a more security-centric approach. Its architecture incorporates advanced encryption and authentication mechanisms. Nevertheless, the very versatility of 5G, serving a plethora of devices from smartphones to IoT sensors, increases its exposure to potential threats. 5G networks also introduce a more decentralized architecture, which can be both an advantage and a challenge. While it allows for localized data processing and reduces the attack surface, it also means threats can be more localized and harder to detect on a global scale. That said, companies such as Verizon or Vodafone have world-class security teams due to their economies of scale compared to most companies who would be deploying Wi-Fi.

What this all boils down to for you, as an architect, is this: ensuring the security of a Wi-Fi-based solution is completely on your shoulders and those of the folks who will be operating your solution going forward. With a 5G-based MEC approach, you are more reliant on the MNO you have partnered with to address this facet.

Capacity

As we discussed in *Chapter 2*, Wi-Fi networks, particularly the widely used Wi-Fi 5 (802.11ac) and the newer Wi-Fi 6 (802.11ax), have made significant strides in increasing capacity with the introduction of technologies such as MU-MIMO. However, as you'll also recall, Wi-Fi **Access Points (APs)** tend to be quite limited in the density they can achieve with these things. While a brute-force approach of deploying a great many APs in response will help to some degree, Wi-Fi will always be inherently limited by its original design parameters that targeted **Local Area Networks** (**LANs**). You can only put so many APs in a single location before the channels/beams interfere with each other.

5G networks offer a substantial leap in capacity compared to predecessors such as 4G/LTE. With the ability to support up to a million devices per square kilometer, 5G is designed to cater to the burgeoning landscape of IoT, smart cities, and densely populated urban areas. Technologies such as beamforming and the use of the millimeter-wave spectrum play a pivotal role in this capacity enhancement. When you factor in narrow-band versions of 5G (NB-IOT), the density achievable is truly staggering compared to Wi-Fi.

Whether this matters to you, as an architect, is completely dependent upon the constellation of application requirements and constraints.

Computer Vision

In many countries, regulations requiring worker protections such as **Personal Protective Equipment** (**PPE**) are published by organizations such as the **Occupational Health and Safety Administration** (**OSHA**) in the US. These regulations commonly require employers to do everything within their power to ensure that PPE is not only available but that it is used at all times in dangerous areas.

The following diagram outlines a solution that involves 5G-capable cameras positioned above the entryways to dangerous areas. These consist of magnetically locked doors that require a worker to scan their badge to unlock the door. The cameras are sending RTSP streams back to GPU-accelerated instances in the AWS Wavelength Zone in the same city, thus giving them a very low RTT – 1 ms in this case:

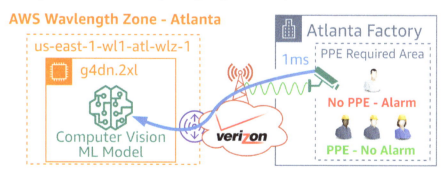

Figure 10.9 – 5G camera applying a CV ML model for PPE detection

The model makes an inference based on what the camera sees regarding whether the worker who just tried to badge in is wearing a hard hat or other PPE appropriate to that area. If this is not true, the door will not unlock even though the user's badge has access to that area otherwise.

The vendor of the building security system owns this application and sells this as a feature to the local factory owners. Because they use 5G-capable cameras and AWS Wavelength, integration is a breeze. There is no need to install local application servers in the factories. Even better, there is no need to deal with the local Wi-Fi or firewalls – both of which are likely inadequate for the task given that these are not office campuses or data centers.

V2X

One of the primary V2X use cases being rolled out now is happening in response to a set of laws passed in both the EU and UK in 2022. As of July 6, 2022, all new cars sold must be fitted with **Intelligent Speed Assist (ISA)** devices, **Autonomous Emergency Braking (AEB)** systems, and data loggers that capture telemetry for analysis by insurers and investigators.

ISA devices

ISA units are devices that work very much like cruise control, except instead of limits being set by the driver, they are able to communicate with local infrastructure to modify the car's throttle in response to road conditions. These could be things such as speed limits, upcoming traffic that the driver can't see yet, icy roads, and so on.

The following diagram shows how an application that manages ISA devices for a given metro area can run in an AWS Wavelength Zone, while other components of the connected vehicle solution run in the parent region on managed services such as AWS IoT FleetWise.

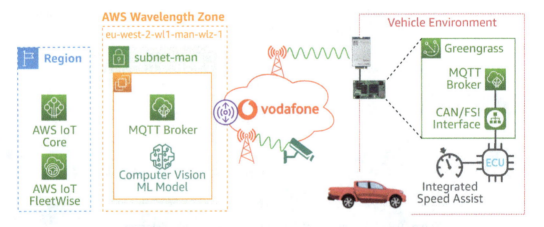

Figure 10.10 – ISA using AWS IoT FleetWise and AWS Wavelength

Notice that the application has a CV ML model that is being informed by multispectrum cameras of the type discussed earlier in this chapter. This lets the application make safety-related inferences about road conditions – for example, it could see "black ice" on the road that the human eye cannot. In response, the application can tell all ISA-equipped vehicles in the area to lower their speed accordingly.

AWS IoT FleetWise

We've already covered AWS IoT Greengrass in previous chapters, but there are other components shown in the preceding diagram that we haven't yet mentioned.

When building any connected vehicle solution, an architect must ask themselves the following questions:

- How can we establish mutual trust between the vehicle and the cloud?
- How can we ingest data from the car's electronic control unit (ECU) to the cloud?
- How can we cache data during periods when the signal is down?
- How can we map these proprietary data structures that vary by make/model/year onto a common namespace?
- How can we make sense of many millions of such data samples over time?
- How can we deploy updates to vehicles in the field?
- How can we discover problems before a customer does?
- How can we respond quickly to individual or fleet-wide issues?
- How can we securely issue commands to the vehicles for control use cases?

AWS' portfolio of IoT services can address these foundational needs, quickly and at scale. See the `Designing Next Generation Vehicle Communication with AWS IoT Core and MQTT` and the `AWS IoT FleetWise Developer Guide` sections of the AWS documentation for detailed information on these services.

Software-defined video production

The sports industry is always looking for ways to reduce video delays to improve the experience for viewers, fans, and players. 5G networks offer a solution with the capability to deliver high bandwidth and reliability, as outlined next:

- **Mixing**: combining various audio and video sources to produce a cohesive program output. Traditionally, large and complex hardware mixers were used to achieve this. These days, virtualized software-based mixers can seamlessly blend live feeds, replay tracks, commentary audio, ambient stadium sounds, overlay graphics, and so on.

- **Switching**: The act of selecting between multiple video feeds. During a live sports event, multiple cameras capture the action from various angles. A video switcher decides, in real time, which camera feed is broadcasted. This determines what the viewer sees at any given moment, whether it's a close-up of a player, an overview of the field, or a replay.

- **Routing**: Directing video and audio signals from their source to their intended destination. This might involve sending a camera feed to a monitor in the control room or routing an audio signal to a commentator's headset. In large-scale sports events, routing becomes complex due to the sheer number of sources and destinations involved.

- **Transcoding**: The process of converting video and audio files from one format to another. With audiences consuming sports content on diverse devices – from high-definition televisions to smartphones – broadcasters need to ensure their content is compatible with all these platforms. Transcoding in real time is crucial for live sports events.

By virtualizing these functions into software running in the cloud, we gain all of the typical advantages any application does. The focus here tends to be on paying only for what you use and having access to the latest hardware/software without constantly having to make large capital purchases for upgrades to specialized equipment.

The problem with doing this historically has been a high sensitivity to latency by these functions. Think about something simple such as synchronization of audio and video feeds. It doesn't take much of a time offset for it to look strange to you as a viewer. This is why even in the era of virtualized video production, crews show up to large sporting events with a portable editing facility. This includes a mini-data center full of rackmount servers in a shipping container towed by an 18-wheeler. Content produced on-site was then sent through expensive satellite uplinks for distribution:

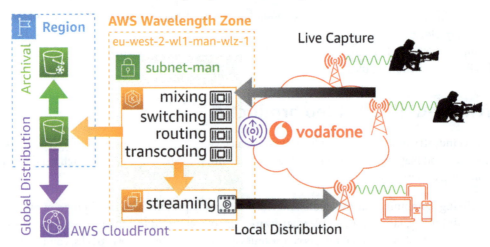

Figure 10.11 – Live video production at the edge with AWS Wavelength

Now, as illustrated in the preceding diagram, 5G-capable cameras and microphones send content straight to the AWS Wavelength Zone – no on-site production facility is needed to maintain the 1-2 ms latency requirement.

There, we have some containers running virtualized video production functions on a set of `g4dn.2xlarge` worker nodes, which enable GPU acceleration of this processing. This is where localized elements such as captions and unique graphics are added. Tasks such as switching and editing are carried out here as well. Content is then either streamed directly to viewers or sent to the cloud for broader distribution through Amazon CloudFront.

To expand on this, fans might opt for personalized viewing options. In-stadium, they could select the camera angles they prefer on their phones and even have AR elements on the video. Additionally, fans attending in person could have their own "fan view" included in the main broadcast.

Summary

In this chapter, we covered a number of 5G-based MEC solutions, delving into the capabilities and potential of AWS Wavelength for public edge solutions, and AWS Private 5G for private implementations. Our exploration of the observability, security, and capacity challenges faced by Wi-Fi networks, as compared to 5G, painted a picture of the strengths and limitations of each technology.

We discussed the capabilities of CV and robotics, empowered by 5G's speed and low latency. The portion covering V2X communication showcased the importance of seamless and rapid communication in enhancing vehicular safety and functionality. The section on software-defined video production shed light on how 5G is reshaping content creation, providing faster, more efficient ways to produce and distribute video.

In the next chapter, we'll dive into how to address the requirements of immersive experiences (AR/VR) using AWS services.

11

Addressing the Requirements of Immersive Experiences with AWS

The ability to deliver high-quality immersive experiences has become a more common expectation across various domains, ranging from gaming and entertainment to professional training and simulation. As the technology underlying these experiences enters its third generation, user demand is rapidly increasing. These applications demand a cohesive integration of visual, auditory, and interactive elements to simulate environments that users find engaging yet simple to adopt.

Building such shared virtual worlds requires significant computing power, low-latency communication, and high-speed storage. Furthermore, development and management tools that can quickly adapt to the new dimensions such applications introduce are needed.

This chapter underscores how AWS plays an instrumental role in meeting the demands of such use cases. We will review how AWS services support robust, scalable, and secure solutions that drive innovation and engagement while minimizing upfront capital investment.

We will cover the following topics:

- Overview of immersive experiences
- Online gaming
- Connected workers
- Workforce development and training
- Augmented reality-enhanced sporting events

Overview of immersive experiences

The term **immersive experiences**, sometimes referred to as **XR**, encompasses **Virtual Reality (VR)**, **Augmented Reality (AR)**, and **Mixed Reality (MR)** use cases. These offer varying levels of immersion and are supported by different underlying technologies and architectural patterns. Such immersive experiences take place in the context of a virtual world, which could be running for the benefit of a single user or shared by dozens or hundreds of users.

The term **metaverse** refers to a future state where these isolated virtual worlds are interconnected into a new worldwide entity. This is analogous to how the internet joined together what were once isolated networks. It became its own overarching entity that no individual or corporation can own, yet is viewed as a single thing in a logical sense. Your ISP sells you internet access, and you expect to be able to access the same things around the world, regardless of who your ISP is. The idea is that, in the future, people will connect to the metaverse, have access to a common set of XR services, and be able to interact with anyone else, regardless of the device or platform they use to connect.

Whether or not this vision plays out in exactly that way, one thing is for certain: XR has become a huge market – and consumers are expecting these kinds of services more and more every day. Worldwide, the XR market is $64 billion (USD), with a combined annual growth rate of 43.5%. This means the market is expected to be over $600 billion in 2028[1].

Virtual Reality (VR)

VR immerses users in a completely simulated environment, shutting out the real world completely. This allows users to experience a completely different reality through purpose-built VR interface equipment such as that shown in the following figure:

---

1   Extended Reality (XR) Market: Global Industry Trends, Share, Size, Growth, Opportunity and Forecast 2023-2028 [https://www.imarcgroup.com/extended-reality-market].

Figure 11.1 – The author building a VR application with an Oculus Quest 2 headset

Of course, VR is great for video games. That said, first-time users tend to be overwhelmed by how many applications for non-gaming purposes are available on app stores such as SteamVR, Pimax, and Viveport.

In VR, you can soar over the Grand Canyon one minute, and explore the Louvre the next. You can learn a new language much quicker by recreating social interactions and connecting what you're saying to virtual objects and situations. Practice that big presentation or an upcoming interview. Attend a concert in another city, watch 3D movies, learn to play the piano, work in a virtual office, or socialize and date in the metaverse. VR is even being used now to prepare for and respond to emergency management situations such as natural disasters, war, or other mass casualty events. AWS is helping with scenarios like this, in both connected and disconnected scenarios.

While these things might have seemed laughable with the earliest versions of VR, that is no longer true. Since about 2020, the quality and performance of devices have dramatically increased as leading VR platforms enter their third generation. At the same time, costs have continued to go down. A high-quality device suitable for use with a wide range of homebound VR use cases can be purchased for a few hundred dollars (USD). Enterprise-grade versions that offer extreme performance or are ruggedized can be had for a few thousand.

During the first generation of VR headsets, only about 3 million devices were purchased in the US – in the third generation, that number has increased tenfold to over 30 million. Worldwide, VR has become a ~$16 billion (USD) market with a 15% annual growth rate (as of 2023).

VR headsets

If you haven't tried a third-generation VR headset, you are missing out. First-time users are invariably impressed with how high-resolution and engaging the virtual world is – it is something that needs to be seen to understand.

While it is very difficult to make absolute comparisons between these devices due to the dozens of differences between any two given models, there are a few key parameters to keep in mind when choosing one:

- **Resolution per eye**: On most headsets, each eye has a little monitor dedicated to it, with a lens in between that makes the image appear much larger on your retina. A higher resolution offers clearer, crisper images, and might well be the parameter that most governs how immersive a VR experience is.

- **Horizontal field of view (H-FOV)**: The number of degrees displayed in the horizontal plane. Most people's eyes have a horizontal field of view somewhere around 190 degrees. While a handful of headsets do have H-FOV this high, most are limited to somewhere around half this value:

Figure 11.2 – Per-eye resolution comparison (from vr-compare.com)

- **Pixel density**: One reason manufacturers tend to limit the FOV is because pixel density is determined by the number of pixels your eye can see *per degree*. A higher pixel density reduces the "screen-door" effect, where grid lines between pixels are visible. A device with a resolution of, say, 1,080 x 720 can increase its pixel density by 50% if it reduces the field of view by a comparable amount. This has the effect of making the image sharper, but reducing immersion slightly because a very narrow FOV can make the user feel like they're wearing blinders. Like most things, you need to find a balance that's most appropriate for your use case.

The following figure shows two devices compared in this dimension:

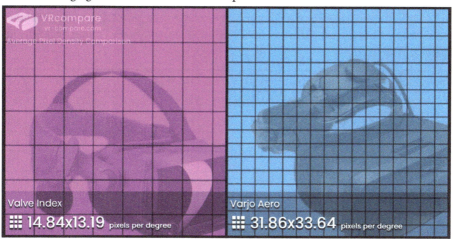

Figure 11.3 – Comparison of pixel density between two VR headset models (from vr-compare.com)

- **Refresh rate**: A higher refresh rate is vital for smooth motion and comfort in the VR environment. A rate of at least 90 Hz will minimize motion sickness and ensure fluid visuals. On most units, you can raise the refresh rate from the default by sacrificing image quality if you are one of those people who experience motion sickness while using a headset.

- **Eye tracking**: A headset with eye tracking can improve the image quality in the specific area you are looking at in the virtual world while reducing the quality of what is in your peripheral vision. Given that peripheral vision is less sharp anyway, this works quite well. This is also known as **Foveated Rendering (FR)**.

- **Standalone or dependent**: About half of the headsets made these days are simple peripherals that attach to a PC or gaming console. This tends to lower the cost of the headset itself, but you still need a reasonably powerful CPU and GPU in the attached PC/console to support high-resolution applications – especially games:

Model Name	Resolution	Refresh	FR	Price
HP Reverb G2	2,160 x 2,160	90 Hz	No	$600
PlayStation VR 2	2,000 x 2,040	120 Hz	Yes	$550
Valve Index	1,440 x 1,600	144 Hz	No	$500
Varjo Aero	2,880 x 2,720	90 Hz	Yes	$990
Arpara VR	2,560 x 2,560	120 Hz	Yes	$600
DPVR E4	1,832 x 1,920	120 Hz	No	$550

Figure 11.4 – Comparison of PC and console-dependent VR headsets

The preceding table gives some examples of headsets that work like this.

The other half or so of headsets sold now actually have their own CPU and GPU embedded within them. This allows them to run a local operating system (typically Android) for independent operation:

Model Name	Resolution	Refresh	FR	CPU	RAM	GPU	Connection	Price
Pico 4	2,160 x 2,160	90 Hz	No	XR2 G1	8	Adreno 650	Wi-Fi	$430
Meta Quest 3	2,064 x 2,208	120 Hz	No	XR2 G2	8	Adreno 740	Wi-Fi	$500
Lenovo TR VRX	2,280 x 2,280	90 Hz	No	XR2+ G1	12	Adreno 650	Wi-Fi	$1,300
Pimax Crystal	2,880 x 2,880	160 Hz	Yes	XR2 G1	8	Adreno 650	Wi-Fi	$1,600
AjnaXR Pro	2,280 x 2,280	90 Hz	Yes	XR2 G1	6	Adreno 650	Wi-Fi	$1,800
Apple Vision Pro	3,400 x 3,400	90 Hz	Yes	Apple M2	16	Apple R1	Wi-Fi	$3,500
XRSpace Manova	1,440 x 1,440	90 Hz	No	845	6	Adreno 630	5G	$500
DPVR P1 Ultra 4K	3,840 x 2,160	90 Hz	No	845	6	Adreno 615	5G	$900
HTC Vive Focus 3	1,920 x 1,920	90 Hz	No	XR2 G1	12	Adreno 650	5G	$1100

Figure 11.5 – Comparison of standalone VR headsets

- The headsets listed in the preceding table are examples of devices that can operate in a completely standalone mode – with no attached PC or gaming console. However, you will notice that the CPU and GPU onboard these devices are meant for mobile devices. This limits the complexity of applications they can support in this mode. That is why even standalone units normally have an attached mode that lets the user leverage a more powerful GPU in a desktop computer.

- **Host connectivity**: Most headsets that can connect to a computer or gaming console offer some form of wired connection for this purpose. This is typically USB 3 or better (5-20 Gbps), although DisplayPort (10-20 Gbps) is used on a few models. Some models offer methods of connecting to their host PC via Bluetooth or Wi-Fi, but in most cases, this results in significantly poorer performance due to the lower throughput and higher latency any wireless technology introduces compared to a direct cable.

- **Internet connectivity**: Headsets attached to a host computer effectively use whatever internet connection that computer has because the application isn't running on the headset. However, headsets operating in fully standalone mode need some way to connect to the internet to download applications and interact with other users online. Almost all can connect to a nearby Wi-Fi network. Some can also connect via 4G/LTE or 5G directly.

- **Accessories**: The design and functionality of any accessories that can be attached to the headset vary considerably. Most headsets include relatively simple hand-held controllers whose position is tracked by cameras on the headset. This limits what a user can do with their hands because they have to grip them – but they are sufficient for games.

  However, they don't work for use cases that require free hand movement or those that need a fine-grained understanding of what each finger is doing. Such applications might require spatial trackers with multiple cameras that visualize the user's hand movements, independent of the headset. Sometimes, these are mounted on the user's wrists, on base stations nearby, or even on special shoes that have upward-facing cameras.

  In some cases, the situation will benefit from haptics – features that give physical feedback to the user. For instance, VR gloves can simulate pressure or resistance when a user touches a virtual object.

Fitness applications are popular in VR marketplaces, and they need a way to determine the user's body position – where are the user's left knee and right foot at this moment? These questions can be answered with cameras on the headset, of course, but that imposes limitations based on line of sight. To help with this, **Inertial Measurement Units** (**IMUs**) can be built into clip-on wearables or even shoes to work around this.

There are also omnidirectional treadmills that let a VR user walk through a virtual environment without going anywhere physically.

Augmented Reality (AR)

In contrast to VR, AR doesn't shut the real world out – it overlays virtual elements onto it. Through devices such as smartphones, AR glasses, or specialized heads-up displays, users can see and interact with virtual elements that appear to exist within their physical environment.

Examples of AR you are probably already familiar with include mobile apps that place furniture in your living room so that you can see how it would look, heads-up displays in cars that overlay directions, and online games such as Pokémon Go:

Figure 11.6 – Examples of devices used with AR

AR glasses

Smart glasses designed for AR have rapidly improved over the past decade. They work similarly to VR headsets but somehow need to allow you to see through them into the real world. This is known as passthrough. Passthrough is sometimes accomplished through the use of transparent lenses that the tiny monitors project onto. Other times, the glasses are opaque, and forward-facing high-resolution cameras are used to show the real world.

Some VR headsets have a passthrough mode via cameras, but it tends to be a secondary function. Passthrough on VR headsets is usually meant to be used to allow the user to temporarily view the real world in low resolution so that they can find a nearby physical object, after which they return to full VR mode. Therefore, be careful when shopping for a VR headset that you want to use as AR glasses.

The technical specifications for AR glasses are similar to those associated with VR headsets. However, because they do not shut out external light sources, the peak brightness (normally measured in nits) is a much larger concern. If the brightness of the projected virtual elements is low, they can be difficult to see when used outdoors.

As these are glasses, they need to be light. Therefore, few models run a local operating system directly in the glasses themselves. They normally offload rendering to the CPU and GPU in a nearby mobile phone or are tethered to their own small computer somewhere nearby (often called a base station or compute pack):

Model Name	Resolution	Refresh	Peak Brightness	Price
Xreal Air 2 Pro	1,920 x 1,080	120 Hz	500	$410
Rokid Max	1,920 x 1,080	120 Hz	600	$440
Nreal Air	1,920 x 1,080	60 Hz	400	$400
Viture One	1,920 x 1,080	60 Hz	1800	$480
Huawei Vision	1,920 x 1,080	60 Hz	480	$430
Microsoft Hololens 2	1,440 x 936	60 Hz	500	$3,500
Google Glass Enterprise 2	640 x 360	60 Hz	300	$1,000

Figure 11.7 – Comparison of AR glasses

Almost all AR glasses support 5G connectivity – if not directly, then through the mobile device they are tethered to.

Mixed Reality (MR)

MR combines aspects of both AR and VR. It not only overlays but also anchors virtual objects to the real world, allowing users to interact with these objects as though they are there. Users can manipulate virtual objects in real time, offering a seamless blend of both realities.

Like AR, high-resolution (or outright transparent) passthrough is required for MR – meaning only a few VR headsets do a good job of supporting MR use cases. Higher-end AR glasses are normally the devices used for MR.

MR is used in medical visualization, remote work, training simulations, and design prototyping.

XR development platforms

Given how complex cross-platform interactive 3D applications are, you probably don't want to write your own real-time rendering engine from scratch. Therefore, the first thing you should do is identify an existing platform/toolset to build with:

*   **3D game engines**: AAA studios have so much expertise in real-time rendering with interactive elements that the most mature VR platforms are 3D gaming engines. These are often the best choices, even if the application you have in mind isn't a game. Examples include `Unreal Engine`, `Unity`, and `CryEngine`.

- **Media production platforms**: Another category has emerged over time from tools that were originally designed for 3D animation in film, television, and similar entertainment media. `Blender` and `Aero` come from this background.

- **Computer-aided design** (**CAD**): These evolved from 3D design tools meant for architects, civil engineers, mechanical engineers, and others. Examples are `Maya` and `Autodesk VRED`.

- **Online XR frameworks**: Some tools were developed from the start to be simplified XR development tools. They are typically easier to get going with but lack support for advanced features available with other types. `Babylon.js` and `O3DE` are popular entries in this category.

The rest of this chapter will focus on how things would be implemented with Unity, which has a good balance between performance, features, and ease of adoption.

Online gaming with Unity

Unity is a cross-platform game engine and comprehensive set of tools developed by Unity Technologies. It supports both 2D and 3D applications on a variety of desktop, mobile, console, and AR/VR platforms. Because it is based on .NET, its primary scripting API is C#.

Unity provides an IDE called the Unity editor. In the Unity editor, you can develop and compile applications that include the Unity runtime for a variety of target devices. Available build targets include Linux, macOS, Windows, Android, iOS, PlayStation 4 and 5, and web servers meant to support browser-based clients via HTML5.

Dedicated servers

A subtype of the Linux, macOS, and Windows build targets is called a `dedicated server`. The term *dedicated* refers to the fact that they serve no other function than to host common components of a shared game – before the advent of this model, one player out of a group would have to host these shared elements on their PC while simultaneously using that same hardware to render their view of the game world. This wasn't ideal as all players had to share the game host's bandwidth, which was typically residential – and worse, the hosting player had an overwhelming performance advantage in PvP games due to their 0 ms RTT.

In addition to maintaining the state of the virtual world, they listen for (and respond to) network requests coming in from many user devices using Unity's **Netcode for GameObjects** (**NGO**) library. This library greatly simplifies the implementation of a game server as it abstracts away the need to worry about things such as low-level networking protocols.

Keep in mind that because no rendering is done on a dedicated server, they do not benefit from GPUs. Also, remember that they are natural integration points for other AWS services. For example, you might want to use S3 to `host common game assets` or DynamoDB to maintain a global state table that spans hundreds of game servers around the world.

Game clients

Game clients are where the player I/O and video rendering happen. That is why, in this model, what dictates game performance the most is the GPU each player has on their setup:

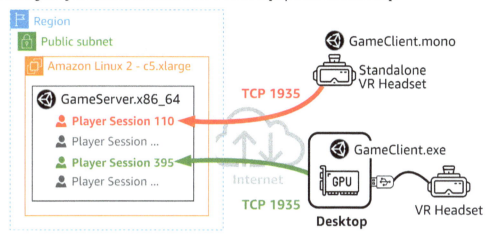

Figure 11.8 – Unity-based shared virtual world using local rendering

Notice that the two players in the preceding diagram are using different builds of the game client. One is using a Windows desktop, while the other is using a headset with its own local Android environment. This is possible because all Unity builds are ultimately .NET. Thus, there is no need for the dedicated server to match the operating system of the game clients that connect to it – or for the game clients to match each other. It is possible to have a Linux-based game server hosting a virtual world for Windows, Android, macOS, and PlayStation clients.

The TCP port of 1935 shown in the preceding diagram is arbitrarily chosen at build time and has no special significance other than to group players using an abstraction that network devices are aware of.

Pixel streaming

Another way to deliver XR for online gaming is via pixel streaming. This works in much the same way as **Virtual Desktop Infrastructure (VDI)** solutions such as Amazon Workspaces do:

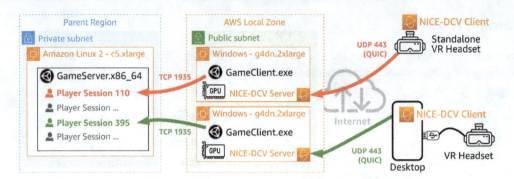

Figure 11.9 – Unity-based shared virtual world using pixel streaming

Notice that in the preceding figure, the client/server configuration remains. Each player is assigned their own EC2 instance running Windows that hosts `GameClient.exe`. Because rendering is now happening on an EC2 instance, the GPUs that are available with certain instance families can be leveraged for offload. Those instances also host a NICE-DCV service that streams the output, which would normally go to an attached display over the internet to NICE-DCV clients. The VR input devices are attached to the server using the `USB Remotization` feature of NICE-DCV:

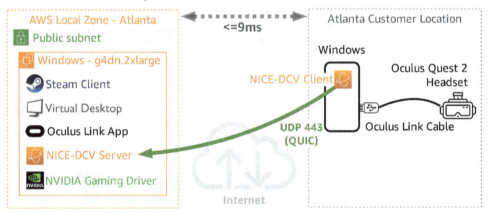

Figure 11.10 – Example of pixel streaming with NICE-DCV to an Oculus Quest 2 VR headset

In the preceding figure, we've zoomed in on the EC2 instance hosting the NICE-DCV server to show the additional components needed in a more specific situation (an Oculus Quest 2 headset streaming a SteamVR application). Note that `Virtual Desktop` is an application available on the SteamVR store:

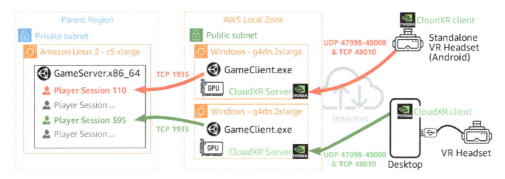

Figure 11.11 – Pixel streaming using the NVIDIA CloudXR SDK

Another option for pixel streaming is the `NVIDIA CloudXR SDK`. As shown in the preceding figure, the architectural model remains the same. However, instead of NICE-DCV, which was developed for general-purpose VDI, we are using NVIDIA's protocol and APIs that were developed specifically for XR. For a bit more clarity, the following table outlines the ports NVIDIA's protocol uses:

Function	Port	Protocol
Control	47999	UDP
Audio	48000	UDP
Video	47998, 48005	UDP
Microphone	48002	UDP
RTSP	48010	TCP

Figure 11.12 – Protocols used by NVIDIA CloudXR Servers

Note that RTSP is only used to establish the connection, at which point all traffic happens over UDP. This reduces the impact of latency, as we discussed in *Chapter 2*.

Advantages

There are significant advantages to this approach, some of which are as follows:

- **Reduced client requirements**: As the game runs on a server, there is less onus on the players to have powerful GPUs.

- **Simplified adoption**: Players can start playing immediately without downloading large game files or installing updates. There is also less dependence on particular device compatibility or user-side setup parameters.

- **Centralized performance management**: Server-side rendering provides consistent performance, independent of the client device's capabilities.

- **Update management**: Admins can update games on the server side, ensuring all players access the latest version without reliance on players downloading patches in a timely fashion.

- **Piracy control**: Since the game isn't downloaded, the risk of piracy is significantly reduced.

Disadvantages

Here are some potential disadvantages of this approach for online gaming:

- **Latency issues**: The time delay between a player's input and the game's response can be noticeable, impacting the gaming experience, especially in fast-paced genres such as first-person shooters. This can be mitigated through the use of AWS Local Zones and AWS Wavelength to push the servers to do this rendering as close to the end users as possible.

- **Bandwidth consumption**: Pixel streaming can use significant bandwidth – particularly in VR games where multiple virtual displays are effectively in use at once. Players with data caps might quickly exhaust their allowances.

- **Connection sensitivity**: User experience is heavily dependent on the player's internet speed and stability. Poor connections result in lower visual quality and increased latency. The primary technical parameter of concern to a player shifts from their GPU to their connectivity.

Keep in mind that access to games of this type is generally sold on VR marketplaces such as SteamVR, Pimax, or Viveport. Therefore, it is simple for the operators of these marketplaces to test a user's configuration and connectivity and warn the player or outright refuse the transaction if their setup is not going to work.

Amazon GameLift plugin for Unity

The `Amazon GameLift plugin for Unity` makes it easier to integrate Amazon GameLift into a Unity-based game. It makes it simple to access Amazon GameLift APIs directly from within your application or use automation services such as AWS CloudFormation to deploy packages that include the EC2 instances needed to host your game servers, along with your binaries.

All multiplayer games need some mechanism for authenticating players, and this can be done using Amazon Cognito. This is a managed service that offloads the need for you to build authentication functionality into your game server binaries directly. The plugin makes this integration easy.

The plugin also gives your game access to digital assets stored in Amazon S3, to kick off AWS Lambda functions, or to output custom server messages for your game sessions to Amazon CloudWatch logs.

The following figure shows an example where a dedicated Linux server binary has been built that leverages multiple such integrations:

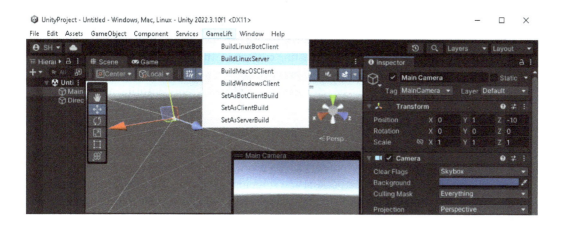

Figure 11.13 – Using the GameLift plugin for Unity to build a dedicated server for Linux

Connected workers

AR has the potential to revolutionize the way connected workers operate across multiple industries by providing them with real-time data, guided instructions, and remote assistance.

Here are some examples:

- **Healthcare**: Doctors can leverage AR glasses for enhanced patient care and surgical precision. During surgeries, they can utilize AR glasses to access and visualize critical patient data, such as vitals, without diverting their attention from the surgical procedure. Furthermore, they can visualize 3D models of the area being operated on, providing them with a detailed view and aiding in precision during the procedure. In another example, nurses making their rounds can receive instant information about a patient's medical history, allergies, and required medications simply by looking at the patient's bed.

- **Emergency response**: AR glasses can dramatically enhance the capabilities of emergency responders and firefighters by providing vital information directly in their line of sight. In a crisis scenario, having hands-free access to building layouts, potential hazards, or the location of individuals in need can significantly enhance rescue efforts. Furthermore, by utilizing AR, these professionals can visualize the safest routes, ascertain structural integrity, or even locate vital resources, such as fire hydrants or medical kits, within their immediate surroundings, thereby improving their situational awareness and operational efficacy.

- **Retail**: Retail professionals equipped with AR glasses can elevate the customer shopping experience by accessing real-time data regarding product availability, specifications, or alternative recommendations. When assisting customers, the glasses can provide immediate insights into inventory levels, pricing, and product location within the store, ensuring accurate and efficient customer service. They can also visualize planograms, ensuring merchandise is displayed optimally.

- **Logistics**: Warehouse workers, when using AR glasses, can optimize picking and storage operations. Visual cues direct workers to the exact location of items, delineate optimal routes within the warehouse, and even confirm pick accuracy, thereby boosting operational efficiency and reducing picking errors. For packing and shipping, AR glasses can superimpose packing instructions and shipping labels, ensuring accurate order fulfillment while mitigating the need for paper instructions or handheld devices.

- **Industrial**: AR can enhance maintenance and repair activities in industrial environments. Technicians wearing AR glasses can visualize step-by-step repair instructions overlaid onto the physical machinery, enabling them to efficiently diagnose issues and execute repairs without needing to reference manuals or schematics. The AR glasses could also provide real-time data on machine performance, upcoming maintenance schedules, or any historical data related to prior breakdowns and repairs.

- **Manufacturing**: These devices can overlay detailed assembly instructions, part identification, and quality-check procedures directly onto the machinery or product being assembled. By providing real-time visual cues, AR glasses reduce the chances of assembly errors, streamline the production process, and ultimately increase manufacturing efficiency. An emerging use case is **collaborative robot** (**cobot**) interaction, where workers receive visual cues and instructions to safely and effectively engage with robotic systems and harmonize human and machine collaboration on the factory floor.

Mobile edge computing (MEC) and connected workers

Let's zoom in on the specific use case of emergency response. There is no opportunity for first responders to ensure an appropriately designed Wi-Fi network is in place before responding to a call. Cellular connectivity is a must in such circumstances.

The `Rokid X-Craft` is an example of AR glasses that are appropriate for first responders. They can be attached to hardhats or fireman's helmets, are waterproof (IP66), and even explosion-proof (Zone 1). Most importantly, they run an Android-based operating system locally that is capable of connecting via 5G directly without needing to be tethered to a mobile device.

The following figure illustrates an architecture for this specific scenario:

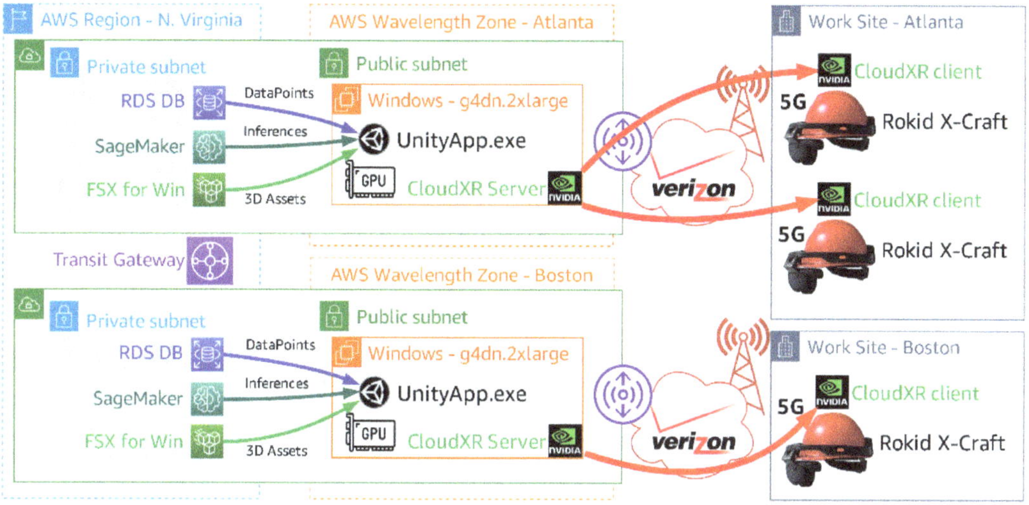

Figure 11.14 – Connected worker application using 5G-capable AR glasses

The glasses offload the heavy lifting of rendering to GPU-accelerated instances in AWS Wavelength Zone, which pixel streams to a lightweight client on the device. 5G supports the high throughput and low latency demands of many dozens of responders concentrated at the site of a disturbance or disaster.

Large databases run in the parent region(s) on Amazon **Relational Database Service (RDS)**. The data points that are kept within them, such as the address of a particular building, combined with incoming data from a local utility can inform ML inferences in AWS SageMaker. These might tell the responder something like "The utility is reporting a gas leak down the street and there is an x% probability it has spread to this location by now."

The library of 3D artifacts for systems like this is typically quite large, so they are kept in-region as well. Amazon FSX for Windows is a managed service that can expose this digital library over SMB 3, the preferred protocol in this case given that the streaming servers in AWS Wavelength are running Windows Server.

Finally, an AWS transit gateway connects the VPCs to provide a path for live communication between responders in cities whose parent Availability Zone is different from theirs.

Workforce development and training

Whether it's a mechanic learning about a new engine type or a medical student observing a simulated surgery, AR provides a significantly more engaging learning environment due to its interactive nature. By overlaying step-by-step instructions, annotations, or even quizzes, these glasses offer hands-on training without the risks associated with real-world mistakes.

Here are some examples of this:

- **Medical**: VR offers a risk-free environment where new surgeons can practice complex surgery without operating on real patients.

- **Manufacturing**: VR can simulate an assembly line environment, allowing workers to practice their tasks, understand the machinery, and familiarize themselves with the production flow.

- **Aerospace and defense**: Pilots and defense personnel require extensive training before they can operate aircraft or sophisticated weaponry. VR simulators can mimic the experience of flying an airplane, helicopter, or even piloting a drone.

- **Customer service**: Retail giants are turning to VR to train their employees in customer service, store management, and inventory handling. Employees can interact with virtual customers, manage virtual stock, and even handle difficult customer scenarios.

- **HR**: AR and VR are not limited to just technical training. Companies are utilizing VR for soft skill training sessions, including leadership programs, communication workshops, and team-building exercises.

- **Automotive**: For mechanics in training, VR can simulate various vehicle issues, allowing budding mechanics to diagnose and address problems in a virtual garage before they work on real cars.

- **Tourism**: For sectors such as tourism and hospitality, AR and VR can offer training on cultural sensitivity, language basics, and even culinary arts. A trainee chef, for instance, could use VR to learn international cooking techniques from a virtual instructor.

AR-enhanced sporting events

AR is now offering fans an enriched viewing experience both on-site and remotely. At the heart of this transformation is the blending of digital information with live action on the field, creating a hybrid environment that immerses spectators in unprecedented ways.

For attendees at the stadium, AR-enabled apps on smartphones or AR glasses can overlay real-time statistics, player bios, and instant replays right onto their field of view. Imagine watching a soccer game and being able to instantly see the stats of a player as they take a free-kick, or viewing a trajectory prediction of a basketball shot as it leaves the player's hand. This layer of real-time information greatly enhances understanding and enjoyment of the nuances of the game.

In the case of remote viewers, AR can recreate a 3D representation of the match, enabling fans to place a miniaturized pitch on their coffee table, for instance, and watch the action unfold from any angle.

Moreover, AR offers innovative advertising opportunities. Traditional static billboards around the pitch or court can be replaced with dynamic, interactive ads that change based on the viewer's location, preferences, or even the ongoing action in the game.

AR-integrated live video

The following figure illustrates an architecture for an AR application that event attendees with mobile devices can use:

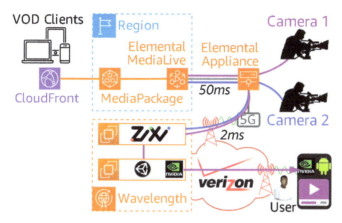

Figure 11.15 – Sporting event attendees viewing camera feeds from angles they can't see

AWS Elemental Live offers a series of on-premises hardware appliances that can natively stream live video to other AWS-managed services in the Elemental family. These services live in-region and can take advantage of the global distribution capabilities of Amazon CloudFront and ML-powered dynamic ad insertion.

At the same time, these appliances can output video streams to other self-managed targets via the Zixi protocol. This is a proprietary UDP-based protocol that has its own error correction and congestion management elements that optimize it for lossless transport of live video streams. In this architecture, the live video streams are split via this mechanism and a 5G hotpot to a Zixi server running in AWS Wavelength.

From there, other instances run a Unity-based AR application that dynamically subscribes to specific camera feeds that the user requests. This is then pixel-streamed to the application with additional graphics and information overlaid. This allows someone watching the event in person to rotate between different camera feeds that have a view of the action that they currently do not. This is particularly desirable with events such as Formula 1 racing, where attendees only see the cars briefly from their vantage point.

Summary

In this chapter, we reviewed the technologies that underpin VR, AR, and MR use cases. They're not just about gaming or entertainment anymore; these technologies, combined with platforms such as Unity, are opening up possibilities in many areas.

We covered how Unity works in the context of online gaming, including elements such as dedicated servers and different types of clients. We reviewed how pixel streaming with Unity through NICE-DCV or CloudXR from NVIDIA opens up VR-based gaming to wider audiences by providing access without the need for high-specification user devices.

Next, we discussed how connected workers can utilize MEC to access crucial data and communication tools via AR glasses, even in remote or hazardous locations. We reviewed how AR and VR have begun revolutionizing workforce training and development, offering simulated, risk-free environments where skills can be honed effectively. Finally, we dove into the application of AR in sporting events, which promises a redefined spectator experience, merging live action with a wealth of digital information and visuals, thus enhancing understanding, engagement, and entertainment value for fans both on-site and remotely.

In the next chapter, we will cover how to turn AWS Snowcone into an IoT gateway.

Part 4:
Implementing Edge Computing Solutions via Hands-On Examples and More

Part Four is a practical guide that shows you how to actually implement some of the architectural concepts introduced in previous chapters. For example, it offers step-by-step instructions on setting up AWS Snowcone as an IoT gateway and its attendant backend services using AWS CloudFormation **Infrastructure-as-Code (IaC)** templates that you can download from this book's Github repository. This part is aimed at applying knowledge to real-world scenarios, demonstrating how to use AWS edge computing services effectively.

This part has the following chapters:

- *Chapter 12, Configuring an AWS Snowcone Device to Be an IOT Gateway*
- *Chapter 13, Deploying a Distributed Edge Computing Application*
- *Chapter 14, Preparing for the Future of Edge Computing with AWS*

12

Configuring an AWS Snowcone Device to Be an IOT Gateway

Now that we've gone over all of the concepts, you're ready to start getting your hands dirty. This chapter will show you step by step how to order an AWS Snowcone device and then configure it as an IoT gateway.

We'll cover the following topics:

- Ordering an AWS Snowcone device
- Deploying the backend in your AWS account
- Preparing the device with the Snowball Edge CLI
- Configuring AWS IoT Greengrass on the Snow device
- Walkthrough of what you've created

Ordering an AWS Snowcone device

This section will walk you through the process of ordering an AWS Snowcone device via the management console.

Check out the following video to view the Code in Action: `https://bit.ly/3Sx3pIf`

Step 1 – Creating an S3 bucket

First, before placing the order, you must create an S3 bucket with a globally unique name.

Create an S3 bucket. We'll be uploading some code there that our automation will reference. Replace `your_bucket_name` with a name that is globally unique.

The command to do this from the CLI is as follows:

```
aws s3 mb s3://your_bucket_name
```

This can also be done from the AWS Management Console under **S3** > **Buckets** > **Create bucket** while in the region you wish to use:

Figure 12.1 – Creating an S3 bucket from the AWS Management Console

If you choose to create it through the AWS Management Console, simply follow the prompts and accept all of the defaults.

Step 2 – Creating a new job

In the AWS Management Console, navigate to your chosen region. Next, search for the **AWS Snow Family** page. Follow that link, then click **Create job** on the left:

Figure 12.2 – Navigating to the AWS Snow Family page

Step 3 – Selecting the job type

Give your job a name and select the **Local compute and storage only** option. Then, click **Next**:

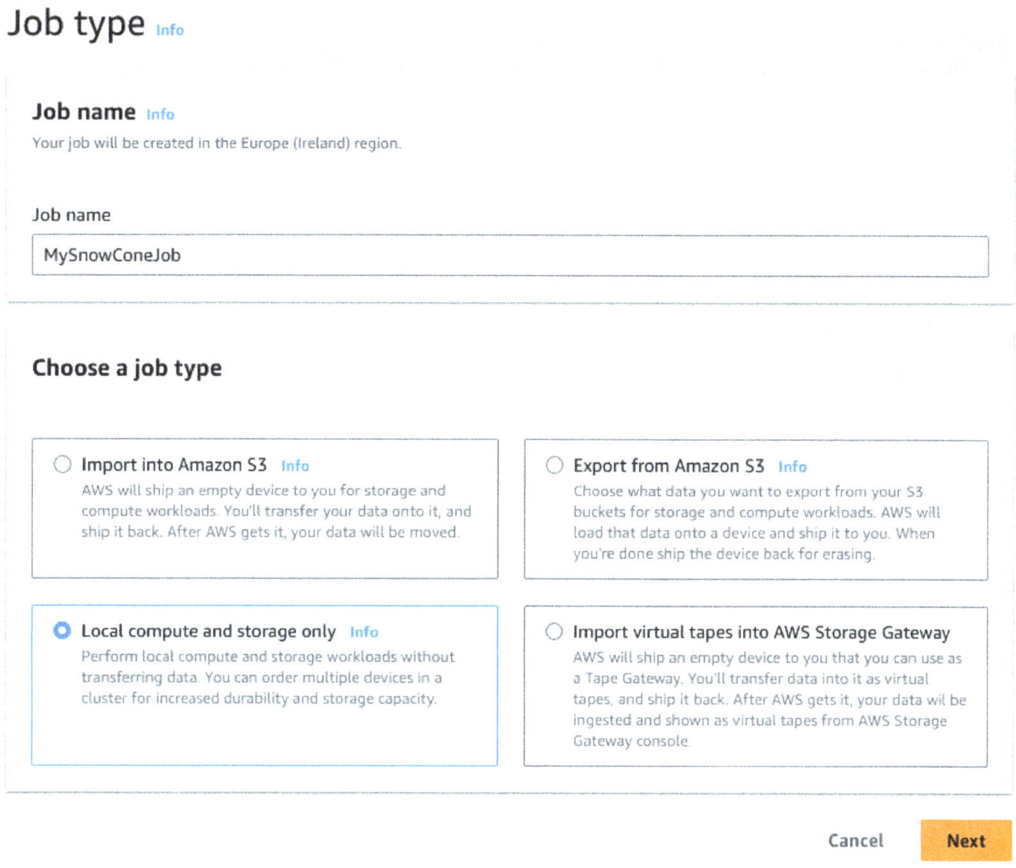

Job type Info

Job name Info
Your job will be created in the Europe (Ireland) region.

Job name

MySnowConeJob

Choose a job type

○ **Import into Amazon S3** Info
AWS will ship an empty device to you for storage and compute workloads. You'll transfer your data onto it, and ship it back. After AWS gets it, your data will be moved.

○ **Export from Amazon S3** Info
Choose what data you want to export from your S3 buckets for storage and compute workloads. AWS will load that data onto a device and ship it to you. When you're done ship the device back for erasing.

● **Local compute and storage only** Info
Perform local compute and storage workloads without transferring data. You can order multiple devices in a cluster for increased durability and storage capacity.

○ **Import virtual tapes into AWS Storage Gateway**
AWS will ship an empty device to you that you can use as a Tape Gateway. You'll transfer data into it as virtual tapes, and ship it back. After AWS gets it, your data will be ingested and shown as virtual tapes from AWS Storage Gateway console.

Cancel   Next

Figure 12.3 – Creating an order job for an AWS Snow Family device

Step 4 – Choosing a form factor

Select the **Snowcone SSD** option:

	Name	Compute	Memory	Storage (HDD)	Stora
○	Snowcone	2 vCPUs	4 GB	8 TB	-
●	Snowcone SSD	2 vCPUs	4 GB	-	14 TE
○	Snowball Edge Storage Optimized with 80TB	40 vCPUs	80 GB	80 TB	1 TB
○	Snowball Edge Compute Optimized	52 vCPUs	208 GB	39.5 TB	7.68
○	Snowball Edge Compute Optimized with GPU	52 vCPUs, GPU	208 GB	39.5 TB	7.68
○	Snowball Edge Compute Optimized	104 vCPUs	416 GB	-	28 TE

Snow devices Info

Figure 12.4 – Selecting the Snowcone SSD option

Check the box for the Amazon Linux 2 AMI:

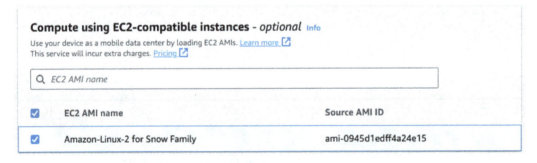

Compute using EC2-compatible instances - *optional* Info

Use your device as a mobile data center by loading EC2 AMIs. Learn more
This service will incur extra charges. Pricing.

🔍 *EC2 AMI name*

☑	EC2 AMI name	Source AMI ID
☑	Amazon-Linux-2 for Snow Family	ami-0945d1edff4a24e15

Figure 12.5 – Selecting the Amazon Linux 2 for Snow Family AMI

Select the S3 bucket you created earlier, then click **Next**:

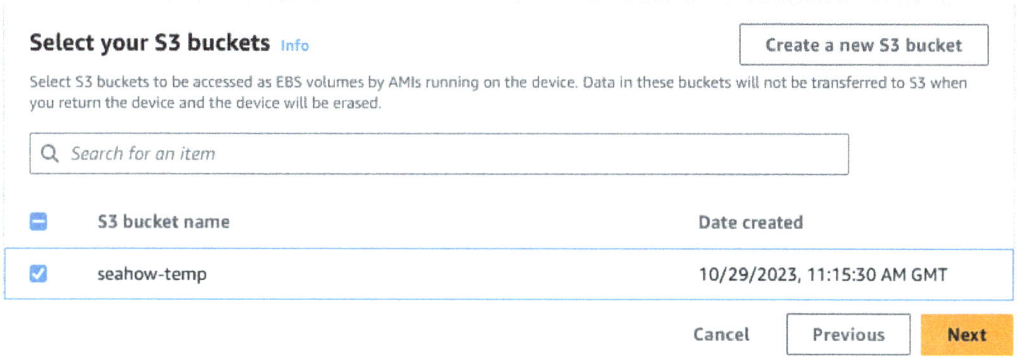

Select your S3 buckets Info   Create a new S3 bucket

Select S3 buckets to be accessed as EBS volumes by AMIs running on the device. Data in these buckets will not be transferred to S3 when you return the device and the device will be erased.

Q Search for an item

	S3 bucket name	Date created
☑	seahow-temp	10/29/2023, 11:15:30 AM GMT

Cancel   Previous   Next

Figure 12.6 – Choosing the S3 bucket created in step 1

Step 5 – Selecting optional components

Check the box labeled **Install AWS IoT Greengrass validated AMI on my Snow device**:

AWS IoT Greengrass

AWS IoT Greengrass for Snow

AWS Snow supports pre-installation of a Greengrass validated AMI on your Snow jobs to enable easy onboarding of IoT workloads for Snow devices. Once you receive the device, you can install AWS IoT Greengrass v2 on this AMI and run your IoT workloads. For more information on getting started with AWS IoT Greengrass for Snow, refer to AWS IoT Greengrass documentation.
This service will incur extra charges. Pricing

☑ Install AWS IoT Greengrass validated AMI on my Snow device

1 service(s) selected

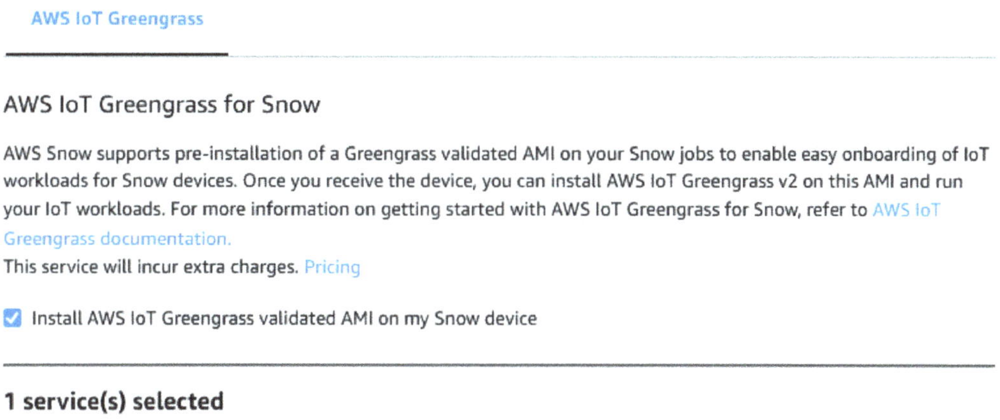

Figure 12.7 – Installing the AWS IoT Greengrass-validated AMI on the Snowcone device

Also, check the boxes labeled **Enable Wireless for Snowcone** and **Manage the device remotely with OpsHub or Snowball Client**. Then, click **Next**:

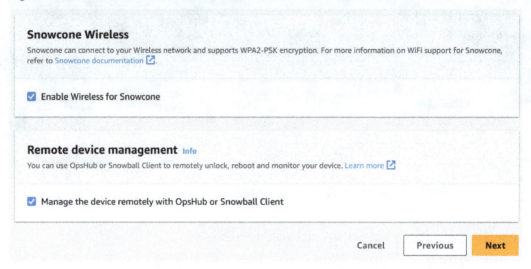

Figure 12.8 – Enabling Wi-Fi and remote management on the device

Step 6 – Setting encryption options

Choose the encryption key you wish to use:

Figure 12.9 – Using the default KMS key

Click the **Create service role** button, and the service will automatically create the needed IAM role for you:

Figure 12.10 – Allowing the job to create a service-linked role for you

Enter your shipping address and choose a delivery speed option:

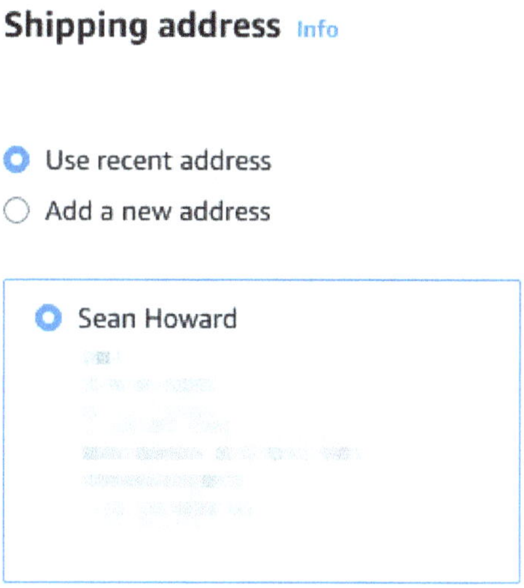

Shipping address Info

○ Use recent address
○ Add a new address

○ Sean Howard

Figure 12.11 – Setting the shipping address

Finally, create a new SNS topic. Give it a name and enter your email address:

Figure 12.12 – Creating a new SNS topic

Click **Next** to proceed.

Step 7 – Confirming your selections

You will now be presented with a summary of your order. Ensure everything is correct, and if it is, click the **Create job** button. You may see an error that the IAM role isn't ready yet if you do this too quickly. Just wait a couple of minutes and try again if this happens:

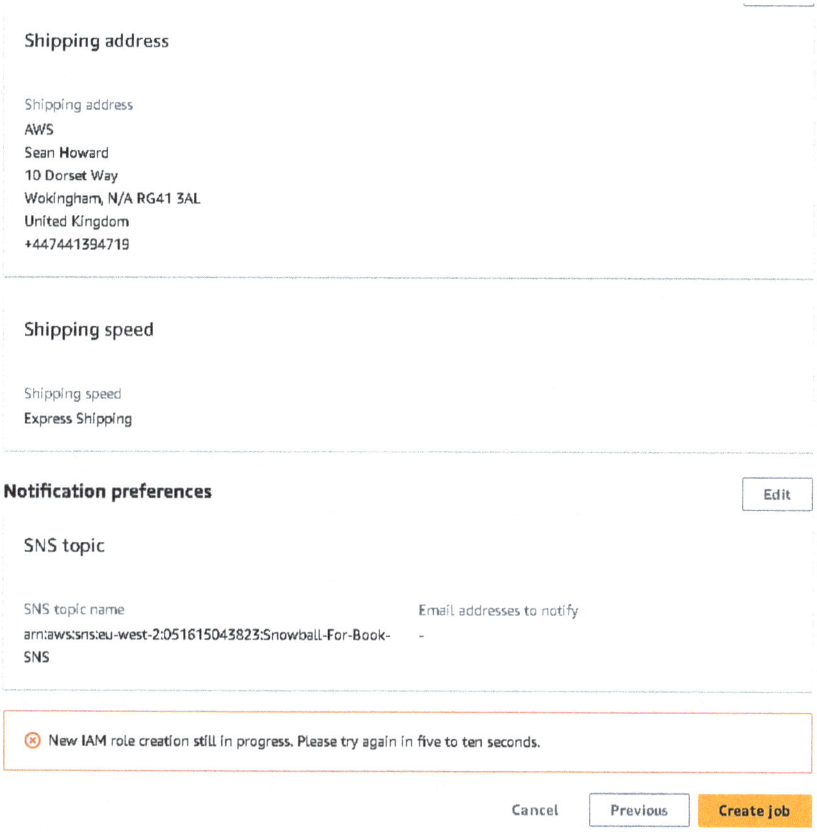

Shipping address

Shipping address
AWS
Sean Howard
10 Dorset Way
Wokingham, N/A RG41 3AL
United Kingdom
+447441394719

Shipping speed

Shipping speed
Express Shipping

Notification preferences   Edit

SNS topic

SNS topic name   Email addresses to notify
arn:aws:sns:eu-west-2:051615043823:Snowball-For-Book-   -
SNS

⊗ New IAM role creation still in progress. Please try again in five to ten seconds.

                                    Cancel   Previous   **Create job**

Figure 12.13 – Confirming the order settings and creating the job

Step 8 – Saving output files

Once the device is shipped, the **Credentials** section of the job details will allow you to copy the unlock code and download the manifest file. Save these to a secure location; they will be needed to unlock the device when it arrives.

Step 9 – Setting up the physical environment

Once the device has been delivered, power it on and connect it to your local network using one of the RJ45 Ethernet ports:

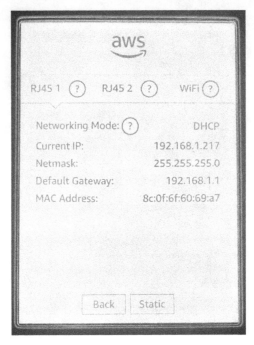

Figure 12.14 – Configuring an Ethernet port for DHCP

DHCP stands for Dynamic Host Configuration Protocol. It is a mechanism used to request an IP address from a pool maintained by a server somewhere on the network. This is usually the the default gateway.

Using the LCD screen, set the IP address for static or DHCP as preferred. Whichever you choose, capture the IP address shown on the screen.

Deploying the backend in your AWS account

AWS IoT Greengrass requires several policies and other constructs to be created in your account before you can attach devices to it. This section will walk you through using a set of AWS CloudFormation templates to automate the creation of these backend components.

> **Recommendation – use Linux or install WSL2**
>
> All command-line operations shown are for a Linux-based operating system. If you are a Windows user, it is recommended you install **Windows Subsystem for Linux 2** (**WSL2**) so that you may directly use the commands shown without needing to translate them to PowerShell or CMD-style batch commands.

Step 1 – Installing and configuring AWS CLI v2

If you do not have AWS CLI v1 installed, simply execute the following commands:

```
curl \
  "https://awscli.amazonaws.com/awscli-exe-linux-x86_64.zip" \
  -o "awscliv2.zip"
unzip awscliv2.zip
sudo ./aws/install
```

If you need to upgrade from AWS CLI v1 to v2, use the following commands instead:

```
curl \
  "https://awscli.amazonaws.com/awscli-exe-linux-x86_64.zip" \
  -o "awscliv2.zip"
unzip awscliv2.zip
sudo ./aws/install \
  --bin-dir /usr/local/bin \
  --install-dir /usr/local/aws-cli \
  --update
```

Whichever method you use, the installation process can take as long as 5 minutes. See the User Guide for Version 2 in the *AWS Command Line Interface* documentation for more details if you run into any trouble.

Next, configure the AWS CLI with administrative credentials for your account, and set the default region. The access key and secret access key can be obtained from the AWS Management Console under **IAM > Users > Credentials**:

```
aws configure \
  set aws_access_key_id "your_access_key"

aws configure \
  set aws_secret_access_key "your_secret_key"

aws configure \
  set region "your_region"
```

To make sure you are connected to your AWS account, execute the following command. It should show your account ID. Here's an example of what you should see:

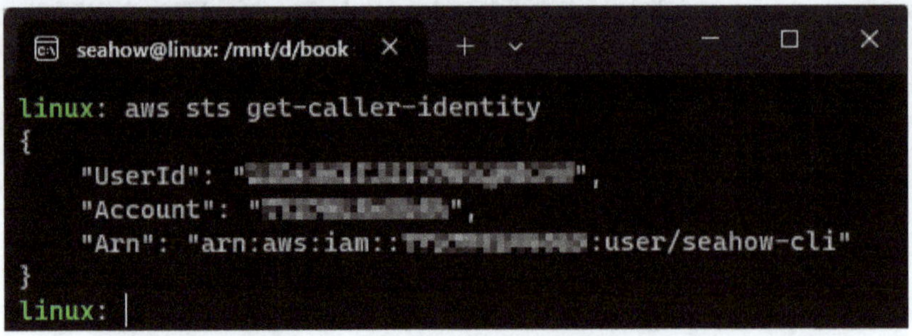

Figure 12.15 – Retrieving the account ID the AWS CLI is connected to

Step 2 – Cloning the code repository from GitHub

Clone the repo from GitHub with the following command:

```
git clone \
  https://github.com/PacktPublishing/Edge-Computing-with-Amazon-Web-
Services.git
```

The next figure gives an example of what you should see:

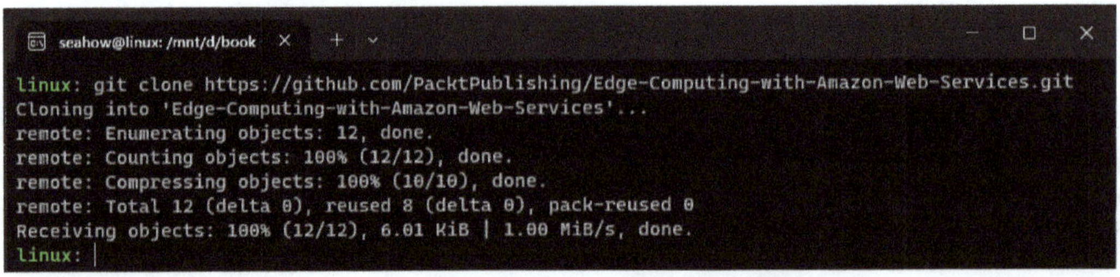

Figure 12.16 – Cloning the repository from GitHub

Step 3 – Copying the code repository to your S3 bucket

We need to copy the contents of the code repo up to your personal S3 bucket so that we can execute CloudFormation templates from there.

Change directory to the root of the code repository you downloaded and synchronize its contents up to the bucket you just created. Replace `your_bucket_name` with whatever you named your S3 bucket:

```
cd Edge-Computing-with-Amazon-Web-Services
aws s3 sync . s3://your_bucket_name
```

The following figure gives an example of what you should see:

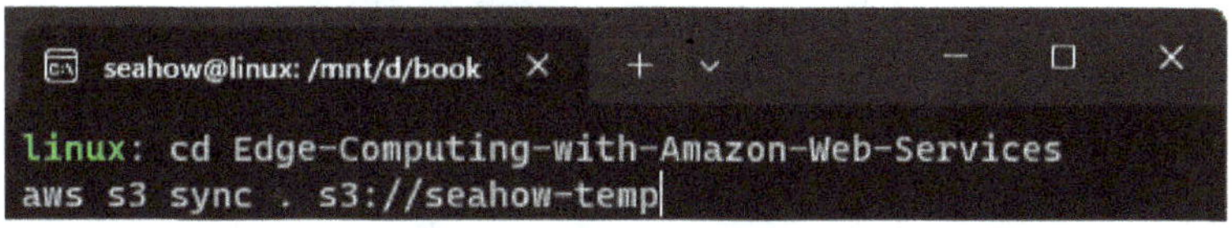

Figure 12.17 – Synchronizing the repository to your S3 bucket

Step 4 – Deploying the CloudFormation templates

Deploy the `main.yaml` CloudFormation template from your bucket. Replace `your_bucket_name` with whatever you named your bucket:

```
aws cloudformation create-stack \
  --stack-name greengrass-backend \
  --template-url https://your_bucket_name.s3.amazonaws.com/greengrass-
backend/cloudformation/main.yaml \
  --capabilities CAPABILITY_NAMED_IAM CAPABILITY_IAM CAPABILITY_AUTO_
EXPAND \
  --parameters ParameterKey=S3Bucket,ParameterValue=your_bucket_name \
ParameterKey=S3Path,ParameterValue=greengrass-backend/cloudformation
```

The following figure shows an example of this:

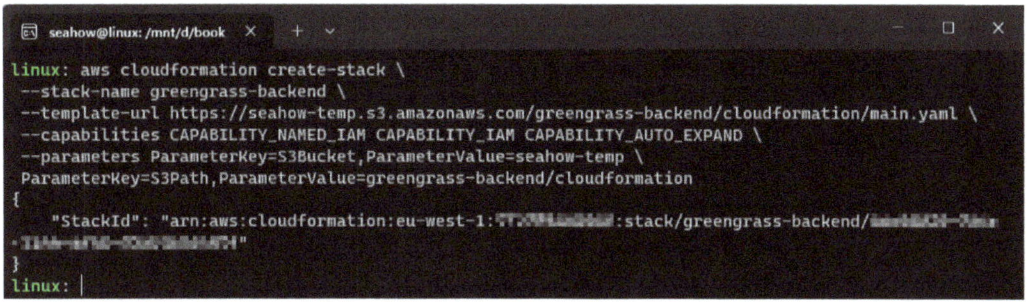

Figure 12.18 – Example of successfully deploying the CloudFormation templates

Wait approximately 5 minutes for the primary stack and its two substacks to fully deploy. You can see their status and troubleshoot any deployment issues in the AWS Management Console under **CloudFormation** > **Stacks**. The following figure shows an example of what they look like after successful deployment:

Figure 12.19 – CloudFormation templates in CREATE_COMPLETE state

Step 5 – Retrieving outputs from CloudFormation and passing them to environment variables

Retrieve the outputs of the Greengrass-backend stack you just deployed and your account ID and insert them into environment variables:

```
export CLAIMPOLICY=$(aws cloudformation describe-stacks \
 --query 'Stacks[?StackName==`greengrass-backend`]
[].Outputs[?OutputKey==`GreengrassProvisioningClaimPolicy`].
OutputValue' \
 --output text)

export SERVICEROLE=$(aws cloudformation describe-stacks \
 --query 'Stacks[?StackName==`greengrass-backend`]
[].Outputs[?OutputKey==`GreengrassServiceRole`].OutputValue' \
 --output text)

export ACCOUNTID=$(aws sts get-caller-identity \
 --query "Account" \
 --output text)
```

Step 6 – Configuring the AWS IoT Greengrass v2 service policy on your account

Associate the AWS IoT Greengrass v2 service policy with the AWS CloudFormation stack generated to your account:

```
aws greengrassv2 associate-service-role-to-account \
--role-arn arn:aws:iam::$ACCOUNTID:role/service-role/$SERVICEROLE
```

The next figure shows an example of this:

Figure 12.20 – Example of successfully associating the AWS IoT Greengrass v2 service role

At this point, everything needed on the backend is created and ready for you to attach your AWS Snowcone device as an AWS IoT Greengrass Core device. The next section will give you the procedure for doing so.

Preparing the device with the Snowball Edge CLI

When a Snow Family device first boots up, it is always locked. You will need the unlock key and manifest file from the preceding section to unlock the device for use.

Step 1 – Installing the AWS Snow CLI client

Download and install the `AWS Snow CLI` client. This is similar to the general-purpose AWS CLI utility but includes commands specific to Snow Family devices:

```
wget https://snowball-client.s3.amazonaws.com/latest/snowball-client-
linux.tar.gz
tar -xvf ./snowball-client-linux.tar.gz
```

> **Note – Setting the build number**
>
> In the following command, make sure you replace `buildnum` with the build number of the client you downloaded. This can be seen by executing the `ls` command.

Now, add the AWS Snow CLI `bin` directory path to your `PATH` variable:

```
export PATH=$PATH:$HOME/snowball-client-linux-buildnum/bin/
```

Figure 12.21 – Adding the AWS Snow CLI bin directory to your PATH variable

More detailed information can be found in the `Using the Snowball Edge Client` section of the *AWS Snow Family* documentation.

Step 2 – Configuring the AWS Snow CLI

You must configure the Snowball Edge client with credentials similar to how you did this with the standard AWS CLI. Don't worry that all the commands start with `snowballEdge` – we use the same utility for all AWS Snow Family devices:

```
snowballEdge configure
```

It will then ask you for the unlock code and path to the manifest file you downloaded earlier. It will also ask you for the default endpoint. This is the IP address of your Snow device that you configured on the LCD panel earlier. It must be formatted with `https://` in front of it, as shown in the following figure:

Figure 12.22 – Configuring the Snowball Edge CLI utility

Step 3 – Unlocking your Snow device with the CLI

Issue the `unlock-device` command:

```
snowballEdge unlock-device
```

This will take a few minutes:

Figure 12.23 – Unlocking the Snow device

Step 4 – Retrieving the output of the unlocking procedure

Check to make sure everything is working properly with the `describe-device` command:

```
snowballEdge describe-device
```

You should receive output like that shown in the following figure:

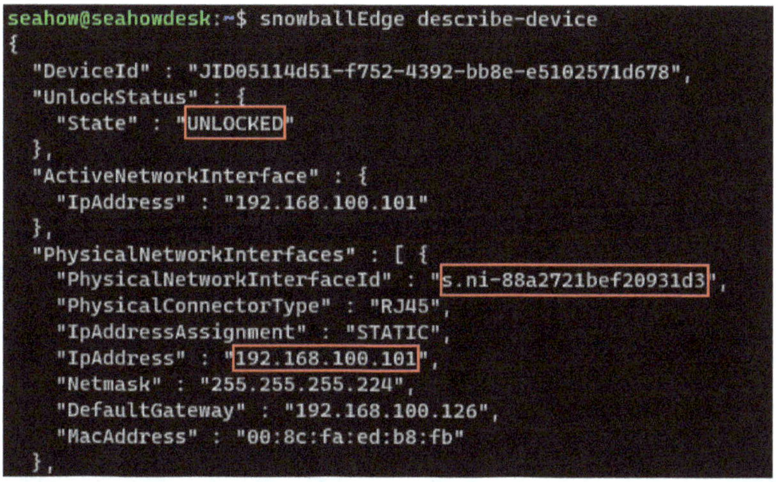

```
seahow@seahowdesk:~$ snowballEdge describe-device
{
  "DeviceId" : "JID05114d51-f752-4392-bb8e-e5102571d678",
  "UnlockStatus" : {
    "State" : "UNLOCKED"
  },
  "ActiveNetworkInterface" : {
    "IpAddress" : "192.168.100.101"
  },
  "PhysicalNetworkInterfaces" : [ {
    "PhysicalNetworkInterfaceId" : "s.ni-88a2721bef20931d3",
    "PhysicalConnectorType" : "RJ45",
    "IpAddressAssignment" : "STATIC",
    "IpAddress" : "192.168.100.101",
    "Netmask" : "255.255.255.224",
    "DefaultGateway" : "192.168.100.126",
    "MacAddress" : "00:8c:fa:ed:b8:fb"
  },
```

Figure 12.24 – Output from the snowballEdge describe-device command

Three important things are highlighted in red:

- `State`: `UNLOCKED`. This means that the process is complete and the device is ready for use.

- `IpAddress`: In this case, it has a value of `192.168.100.101`, which is what we set this particular RJ45 interface to statically from the LCD panel. This refers to the IP that the Snow Family device itself is using for management and routing tasks (where applicable). Any EC2 instances we start on this interface will get their own separate IP and MAC address.

- `PhysicalNetworkInterfaceId`: Copy this value, as we will need it in subsequent steps so that we can start EC2 instances attached to this interface. Alternatively, you can select a different RJ45 interface or even a Wi-Fi interface for this purpose if you wish.

Step 5 – Creating a virtual network interface

Before we launch our EC2 instance to host the AWS IoT Greengrass v2 agent, we should create a **Virtual Network Interface** (VNI) as follows. Replace `your_ip` and `your_netmask` with appropriate values for your network:

```
snowballEdge create-virtual-network-interface \
  --physical-network-interface-id your_pni_id \
  --ip-address-assignment STATIC \
  --static-ip-address-configuration IpAddress=your_ip,Netmask=your_
netmask
```

In the next example, we've chosen to create a VNI with a static IP assigned to it. VNIs are created separately from EC2 instances because they can be detached/reattached as you launch and terminate instances. Copy the `VirtualNetworkInterfaceArn` value as you will need it in subsequent steps:

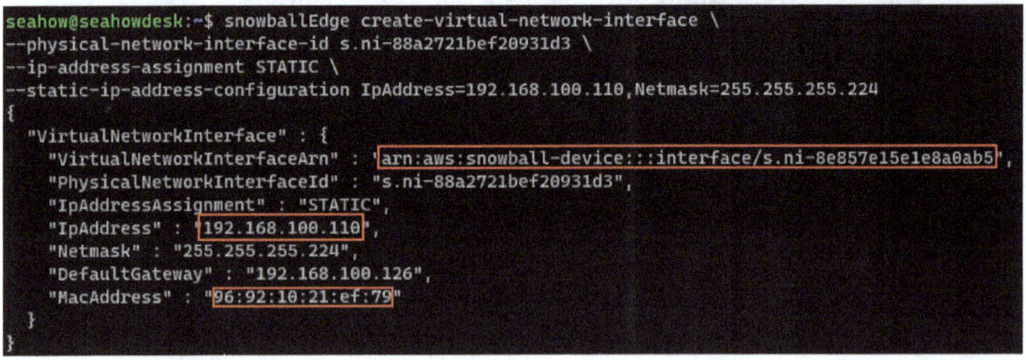

```
seahow@seahowdesk:~$ snowballEdge create-virtual-network-interface \
--physical-network-interface-id s.ni-88a2721bef20931d3 \
--ip-address-assignment STATIC \
--static-ip-address-configuration IpAddress=192.168.100.110,Netmask=255.255.255.224
{
  "VirtualNetworkInterface" : {
    "VirtualNetworkInterfaceArn" : "arn:aws:snowball-device:::interface/s.ni-8e857e15e1e8a0ab5",
    "PhysicalNetworkInterfaceId" : "s.ni-88a2721bef20931d3",
    "IpAddressAssignment" : "STATIC",
    "IpAddress" : "192.168.100.110",
    "Netmask" : "255.255.255.224",
    "DefaultGateway" : "192.168.100.126",
    "MacAddress" : "06:92:10:21:ef:79"
  }
}
```

Figure 12.25 – Output from the create-virtual-network-interface command

Step 6 – Obtaining local credentials from your Snow device

We can use the standard AWS CLI client for EC2 instance operations on a Snow Family device, but we must first obtain a special set of local credentials from the device.

First, use the `list-access-keys` command:

```
snowballEdge list-access-keys
```

Using the output of the `list-access-keys` command, issue the `get-secret-access-key` command:

```
snowballEdge get-secret-access-key --access-key-id your_key
```

The following figure shows an example of this:

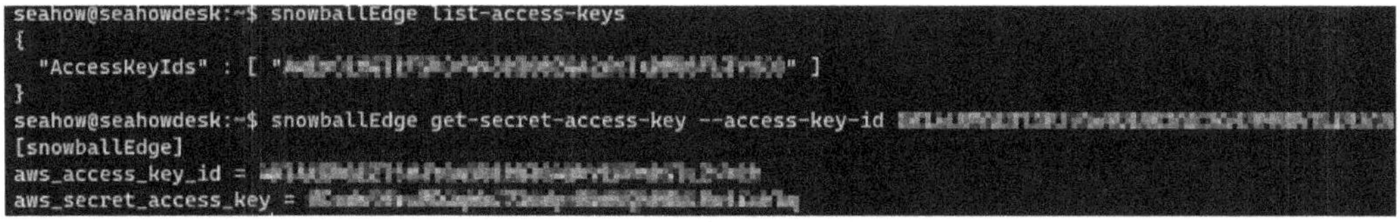

```
seahow@seahowdesk:~$ snowballEdge list-access-keys
{
  "AccessKeyIds" : [ "▒▒▒▒▒▒▒▒▒▒▒▒▒▒▒▒▒▒" ]
}
seahow@seahowdesk:~$ snowballEdge get-secret-access-key --access-key-id ▒▒▒▒▒▒▒▒▒▒▒▒▒▒▒▒▒▒
[snowballEdge]
aws_access_key_id = ▒▒▒▒▒▒▒▒▒▒▒▒▒▒▒▒▒▒
aws_secret_access_key = ▒▒▒▒▒▒▒▒▒▒▒▒▒▒▒▒▒▒
```

Figure 12.26 – Obtaining local device credentials

You can copy all three lines that this command issues into your ~/.aws/credentials file. This will create a profile called snowballEdge. Alternatively, you can run the aws configure command to create the profile interactively:

```
aws configure --profile snowballEdge
```

Regardless of the method chosen, you should set the default region to snow.

Step 7 – Retrieving the certificate from your Snow device

Now, we need to download the Snow Family device's certificate. This is how we make sure any commands we issue with the AWS CLI are encrypted in flight.

First, use the list-certificates command. Copy the ARN of the certificate from the output:

```
snowballEdge list-certificates
```

Second, use the get-certificate command with the ARN you just copied and redirect its output to a file called ca-bundle.pem:

```
snowballEdge get-certificate --certificate-arn your_arn > ca-bundle.
pem
```

Finally, configure the snowballEdge profile you created in the AWS CLI to always use this certificate:

```
aws configure set profile.snowballEdge.ca_bundle ./ca-bundle.pem
```

The following figure shows an example of this process:

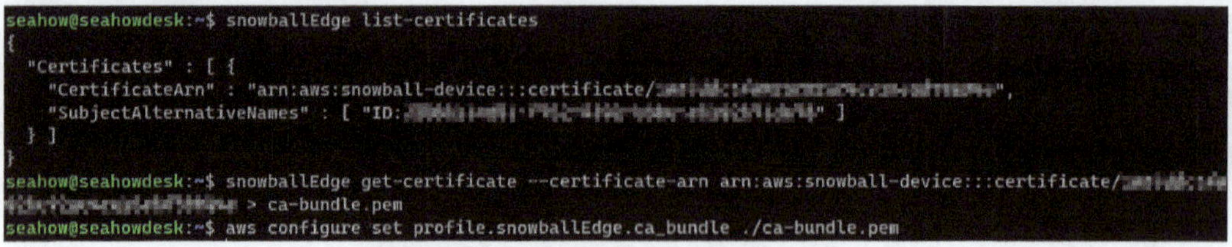

```
seahow@seahowdesk:~$ snowballEdge list-certificates
{
  "Certificates" : [ {
    "CertificateArn" : "arn:aws:snowball-device:::certificate/████████████████████████",
    "SubjectAlternativeNames" : [ "ID:███████████████████████████████" ]
  } ]
}
seahow@seahowdesk:~$ snowballEdge get-certificate --certificate-arn arn:aws:snowball-device:::certificate/████████
████████████████ > ca-bundle.pem
seahow@seahowdesk:~$ aws configure set profile.snowballEdge.ca_bundle ./ca-bundle.pem
```

Figure 12.27 – Configuring the snowballEdge profile in AWS CLI to use the device's certificate

Step 8 (optional) – Adding the EC2 endpoint URL to your AWS CLI config file

This step is optional but recommended as it saves you from needing to remember to tell AWS CLI commands what the endpoint IP address is every time you use a command against your Snow Family device.

Using your favorite text editor (such as vi or nano), open your AWS CLI config file (usually ~/.aws/config) and make sure your snowballEdge profile section looks like this:

```
[profile snowballEdge]
ca_bundle = ./ca-bundle.pem
region = snow
services = snowballEdge

[services snowballEdge]
ec2 =
  endpoint_url = https://your_snow_ip:8243
```

Notice the last five lines. There is a services = snowballEdge reference under the profile pointing to a separate section that actually contains the IPs for your endpoint definitions:

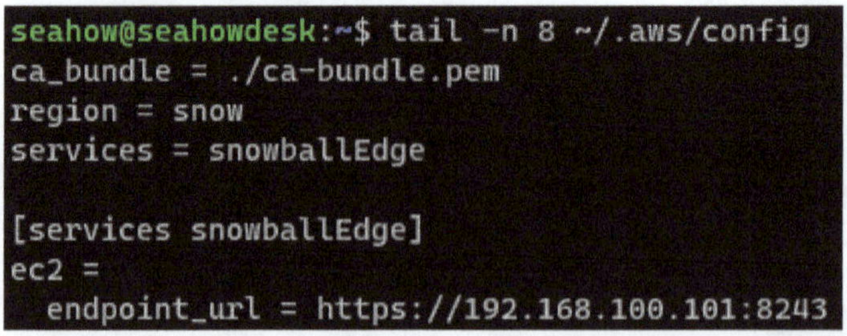

```
seahow@seahowdesk:~$ tail -n 8 ~/.aws/config
ca_bundle = ./ca-bundle.pem
region = snow
services = snowballEdge

[services snowballEdge]
ec2 =
  endpoint_url = https://192.168.100.101:8243
```

Figure 12.28 – AWS CLI config file with EC2 endpoint address specified

We are now ready to run AWS CLI commands against the Snow device for EC2 instance operations.

Configuring AWS IoT Greengrass on the Snow device

The AWS IoT Greengrass agent can technically run from any computer. It doesn't matter if it's a physical or virtual machine, and Greengrass supports a wide range of operating systems. That said, for this exercise, we will be using the AWS IoT Greengrass-validated AMI that you selected to be installed when you ordered the device.

Step 1 – Creating an EC2 keypair on the Snow device

The very first task before launching any EC2 instance is to create a key pair. In this case, we will do it from the CLI and output its contents to a local file. We also want to set the permissions to 600 on the my-keypair.pem file or SSH will fail later:

```
aws ec2 create-key-pair \
    --key-name my-keypair \
    --key-type rsa \
    --key-format pem \
    --query "KeyMaterial" \
    --output text > my-keypair.pem \
    --profile snowballEdge


chmod 600 my-keypair.pem
```

Step 2 – Obtaining the ImageId value from your Snow device

Next, we need to know the ImageId value of the AWS IoT Greengrass-validated AMI on our Snow family device to be able to launch it. This can be obtained with the AWS CLI using the describe-images command for our snowballEdge profile:

```
aws ec2 describe-images --profile snowballEdge
```

Copy the ImageId value from the appropriate AMI, as shown in the following figure:

```
seahow@seahowdesk:~$ aws ec2 describe-images --profile snowballEdge
{
    "Images": [
        {
            "ImageId": "s.ami-8276f10eaa69e8ec4",
            "Public": false,
            "State": "AVAILABLE",
```

Figure 12.29 – Using the AWS CLI to describe available AMIs on a Snow Family device

Step 3 – Launching the EC2 instance on your Snow device

Launch the AWS IoT Greengrass-validated AMI using the `ImageId` value from *step 1*:

```
aws ec2 run-instances \
    --image-id your_image \
    --key-name my-keypair \
    --profile snowballEdge
```

Copy the `InstanceId` value from the output, as highlighted in the next figure:

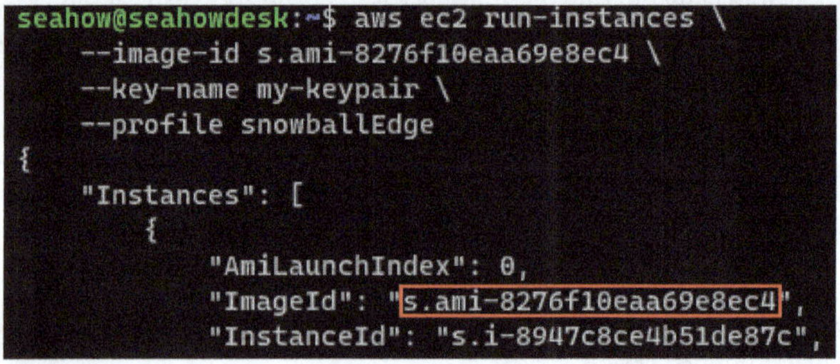

Figure 12.30 – Launching an EC2 instance on a Snow Family device

Step 4 – Attaching the VNI to your EC2 instance

Attach the VNI with a static IP that you created earlier to this newly running instance:

```
aws ec2 associate-address \
    --instance-id your_id \
    --public-ip your_ip \
    --profile snowballEdge
```

It might seem strange telling it to use a public IP, but remember – in the context of Snow Family devices, public IPs are on your local network (which may well be private):

Figure 12.31 – Associating a VNI with an EC2 instance

Step 5 – SSHing into your EC2 instance

Now, you are ready to SSH into the EC2 instance running on your Snow Family device:

```
ssh -i ./my-keypair.pem ec2-user@192.168.100.110
```

You will be asked whether to add this host to the list of known hosts, as shown in the next figure. Answer yes, and you will be connected:

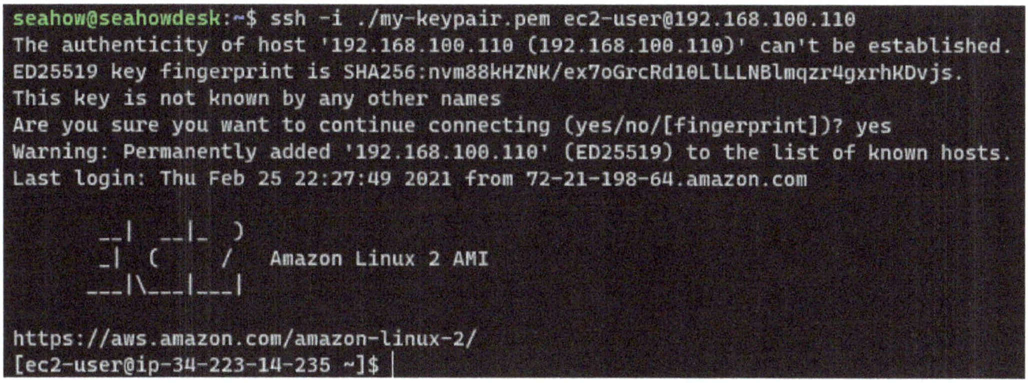

```
seahow@seahowdesk:~$ ssh -i ./my-keypair.pem ec2-user@192.168.100.110
The authenticity of host '192.168.100.110 (192.168.100.110)' can't be established.
ED25519 key fingerprint is SHA256:nvm88kHZNK/ex7oGrcRd10LlLLNBlmqzr4gxrhKDvjs.
This key is not known by any other names
Are you sure you want to continue connecting (yes/no/[fingerprint])? yes
Warning: Permanently added '192.168.100.110' (ED25519) to the list of known hosts.
Last login: Thu Feb 25 22:27:49 2021 from 72-21-198-64.amazon.com


       __|  __|_  )
       _|  (     /   Amazon Linux 2 AMI
      ___|\___|___|

https://aws.amazon.com/amazon-linux-2/
[ec2-user@ip-34-223-14-235 ~]$ |
```

Figure 12.32 – Connecting to the AWS IoT Greengrass EC2 instance on a Snow Family device

Step 6 – Installing the AWS IoT Greengrass prerequisites into your EC2 instance

The following command string will install the prerequisites needed to run the AWS IoT Greengrass v2 agent. These include AWS CLI v2, Python 3, and Java 8:

```
curl "https://awscli.amazonaws.com/awscli-exe-linux-x86_64.zip" \
  -o "awscliv2.zip" && \
  unzip awscliv2.zip && \
  sudo ./aws/install && \
  sudo yum -y install python3 java-1.8.0-openjdk
```

Grant the root user permission to run the AWS IoT Greengrass v2 agent:

```
sudo sed -in 's/root\tALL=(ALL)/root\tALL=(ALL:ALL)/' /etc/sudoers
```

Step 7 – Installing the AWS IoT Greengrass v2 agent onto your EC2 instance

Download the AWS IoT Greengrass v2 agent software:

```
curl -s https://d2s8p88vqu9w66.cloudfront.net/releases/greengrass-
nucleus-latest.zip > greengrass-nucleus-latest.zip
unzip greengrass-nucleus-latest.zip -d GreengrassCore
rm greengrass-nucleus-latest.zip
```

Temporarily grant the AWS CLI credentials to your AWS account.

> **Note – do not use the credentials you obtained from the Snow device**
>
> Use the same credentials you used to deploy the CloudFormation templates containing the backend components.

Replace `your_key`, `your_secret_key`, and `your_region` as appropriate in the following commands:

```
export AWS_ACCESS_KEY_ID=your_key
export AWS_SECRET_ACCESS_KEY=your_secret_key
export AWS_REGION=your_region
```

Now, we need to retrieve some values from the backend components we deployed earlier, as they are dynamically named:

```
export DEVICEPOLICY=$(aws iot list-policies \
 --query 'policies[?starts_with(policyName, `greengrass-device-
default-policy`) == `true`].policyName' \
 --output text)

export TOKENEXCHANGEROLE=$(aws iam list-roles \
 --query 'Roles[?starts_with(RoleName, `greengrass-token-exchange-
role`) == `true`].RoleName' \
 --output text)

export TOKENEXCHANGEROLEALIAS="$TOKENEXCHANGEROLE-alias"

export GGC_THING_NAME="mysnowcone"
```

```
export GGC_THING_GROUP=$(aws iot list-thing-groups \
  --query 'thingGroups[?starts_with(groupName, `thing-group-`) ==
`true`].groupName' \
  --output text)
```

Exporting key variables prior to running the AWS IoT Greengrass v2 installer:

```
[ec2-user@ip-34-223-14-235 ~]$ export DEVICEPOLICY=$(aws iot list-policies \
>   --query 'policies[?starts_with(policyName, 'greengrass-device-default-policy') == 'true'].policyName' \
>   --output text)
[ec2-user@ip-34-223-14-235 ~]$
[ec2-user@ip-34-223-14-235 ~]$ export TOKENEXCHANGEROLE=$(aws iam list-roles \
>   --query 'Roles[?starts_with(RoleName, 'greengrass-token-exchange-role') == 'true'].RoleName' \
>   --output text)
[ec2-user@ip-34-223-14-235 ~]$
[ec2-user@ip-34-223-14-235 ~]$ export TOKENEXCHANGEROLEALIAS="$TOKENEXCHANGEROLE-alias"
[ec2-user@ip-34-223-14-235 ~]$
[ec2-user@ip-34-223-14-235 ~]$ export GGC_THING_NAME="mysnowcone"
[ec2-user@ip-34-223-14-235 ~]$
[ec2-user@ip-34-223-14-235 ~]$ export GGC_THING_GROUP=$(aws iot list-thing-groups \
>   --query 'thingGroups[?starts_with(groupName, 'thing-group-') == 'true'].groupName' \
>   --output text)
[ec2-user@ip-34-223-14-235 ~]$
```

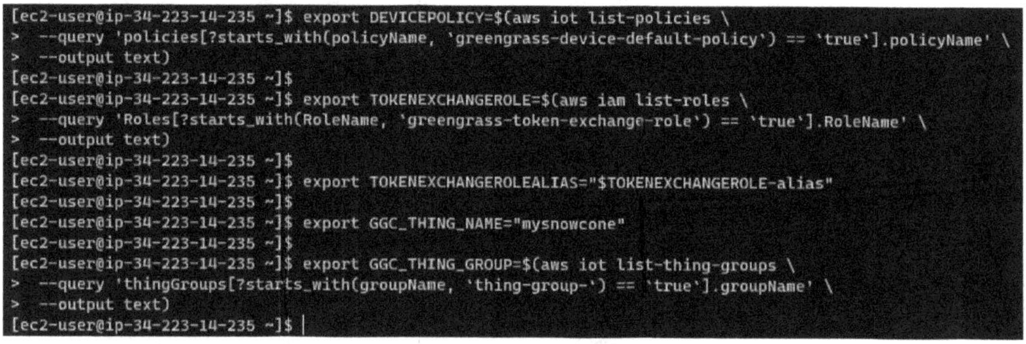

Figure 12.33 – Exporting key variables

Now, we're ready to execute the AWS IoT Greengrass v2 installer:

```
sudo -E java -Droot="/greengrass/v2" -Dlog.store=FILE \
    -jar ./GreengrassCore/lib/Greengrass.jar \
    --aws-region $AWS_REGION \
    --thing-name $GGC_THING_NAME \
    --thing-group-name $GGC_THING_GROUP \
    --thing-policy-name $DEVICEPOLICY \
    --tes-role-name $TOKENEXCHANGEROLE \
    --tes-role-alias-name $TOKENEXCHANGEROLEALIAS \
    --component-default-user ggc_user:ggc_group \
    --provision true \
    --setup-system-service true \
    --deploy-dev-tools true
```

An example of what the output should look like is shown in the next figure:

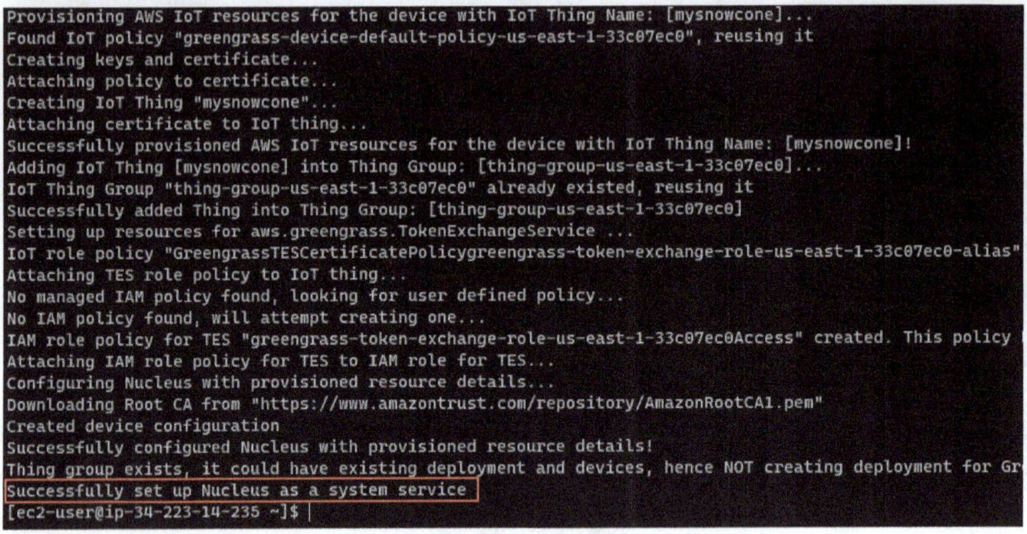

```
Provisioning AWS IoT resources for the device with IoT Thing Name: [mysnowcone]...
Found IoT policy "greengrass-device-default-policy-us-east-1-33c07ec0", reusing it
Creating keys and certificate...
Attaching policy to certificate...
Creating IoT Thing "mysnowcone"...
Attaching certificate to IoT thing...
Successfully provisioned AWS IoT resources for the device with IoT Thing Name: [mysnowcone]!
Adding IoT Thing [mysnowcone] into Thing Group: [thing-group-us-east-1-33c07ec0]...
IoT Thing Group "thing-group-us-east-1-33c07ec0" already existed, reusing it
Successfully added Thing into Thing Group: [thing-group-us-east-1-33c07ec0]
Setting up resources for aws.greengrass.TokenExchangeService ...
IoT role policy "GreengrassTESCertificatePolicygreengrass-token-exchange-role-us-east-1-33c07ec0-alias"
Attaching TES role policy to IoT thing...
No managed IAM policy found, looking for user defined policy...
No IAM policy found, will attempt creating one...
IAM role policy for TES "greengrass-token-exchange-role-us-east-1-33c07ec0Access" created. This policy
Attaching IAM role policy for TES to IAM role for TES...
Configuring Nucleus with provisioned resource details...
Downloading Root CA from "https://www.amazontrust.com/repository/AmazonRootCA1.pem"
Created device configuration
Successfully configured Nucleus with provisioned resource details!
Thing group exists, it could have existing deployment and devices, hence NOT creating deployment for Gr
Successfully set up Nucleus as a system service
[ec2-user@ip-34-223-14-235 ~]$ |
```

Figure 12.34 – Successful installation of the AWS IoT Greengrass v2 agent

You are now finished with the configuration of the Snow Family device itself. Disconnect your SSH session.

Walkthrough of what you've created

Now that your AWS Snowcone device is configured as an AWS IoT Greengrass v2 Core device, let's go over what that actually means.

Controlling your EC2 instance remotely

AWS IoT offers a feature called Secure Tunnel that allows you to remotely connect to a device anywhere in the world from the AWS Management Console. As long as the AWS IoT Greengrass v2 agent can reach AWS over the internet, you will be able to connect.

Log in to the AWS Management Console and navigate to **IoT Core** > **Manage** > **All devices** > **Things** > Your Snowcone name. Click on **Create secure tunnel**, as shown in the next figure:

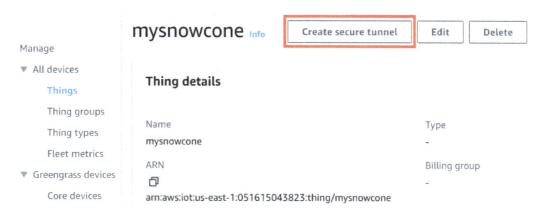

Figure 12.35 – Initiating the creation of a secure tunnel

Accept the defaults to use the **Quick setup** method, and click **Next**:

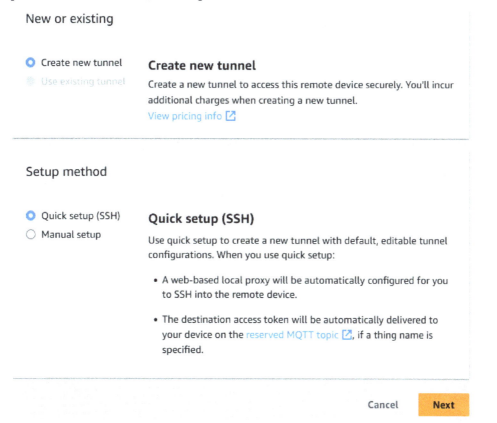

Figure 12.36 – Using the Quick setup method

After this, you will be asked to confirm once again and offered the opportunity to download access tokens. This is not necessary for what we want to do, so simply click **Done**.

At this point, you will be taken to the detail page of the secure tunnel. In the lower right of the screen, under SSH, click **Connect**, as shown in the next figure:

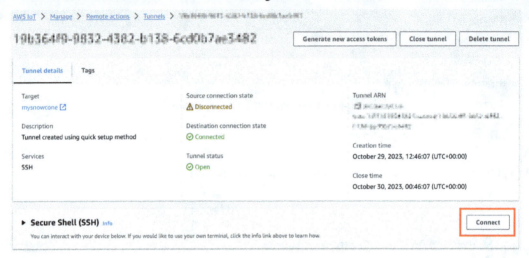

Figure 12.37 – Connecting to the secure tunnel

You will then see a popup asking how you wish to authenticate. Use the same method you would if you were connecting normally over SSH. Set the username to ec2-user and upload the EC2 keypair you created earlier when you launched the instance. An example is shown in the next figure:

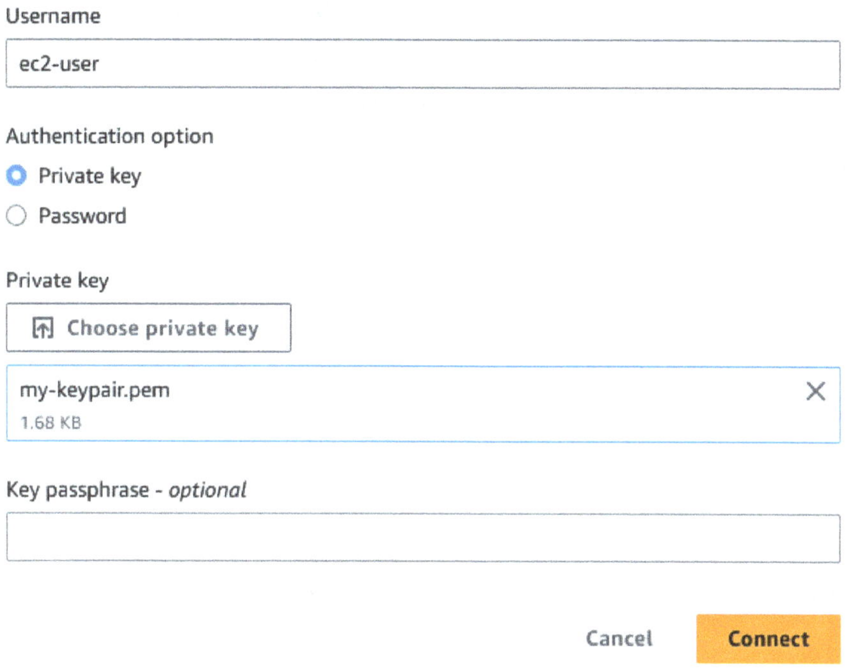

Username

ec2-user

Authentication option
- ● Private key
- ○ Password

Private key

⬆ Choose private key

my-keypair.pem   ✕
1.68 KB

Key passphrase - *optional*

Cancel   **Connect**

Figure 12.38 – Authenticating to the secure tunnel

Depending upon the latency of your internet connection, this may take a moment. Once it is finished, you will be able to interact with the EC2 instance you deployed onto the device from within the AWS Management Console. This is shown in the following figure:

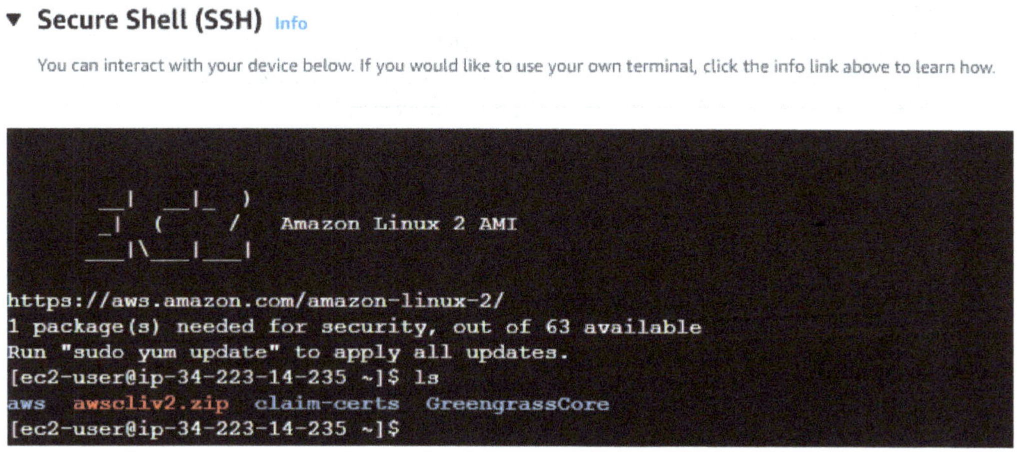

▼ **Secure Shell (SSH)** Info

You can interact with your device below. If you would like to use your own terminal, click the info link above to learn how.

```
   _|  _| _ )
   _|  (     /     Amazon Linux 2 AMI
  ___|\___|___|

https://aws.amazon.com/amazon-linux-2/
1 package(s) needed for security, out of 63 available
Run "sudo yum update" to apply all updates.
[ec2-user@ip-34-223-14-235 ~]$ ls
aws  awscliv2.zip  claim-certs  GreengrassCore
[ec2-user@ip-34-223-14-235 ~]$
```

Figure 12.39 – Connecting to SSH from within the AWS Management Console

Exploring the Greengrass components on your device

The CloudFormation templates you deployed earlier included a rather complex configuration that was assigned to your EC2 instance the moment it connected to the Greengrass service. Let's take a look at what this consists of as a jumping-off point for further exploration of what these components can do for you.

In the AWS Management Console, navigate to **AWS IoT Core** > **Manage** > **Greengrass devices** > **Core devices** and then click on your Snowcone name.

This will take you to an overview page, as shown in the following figure:

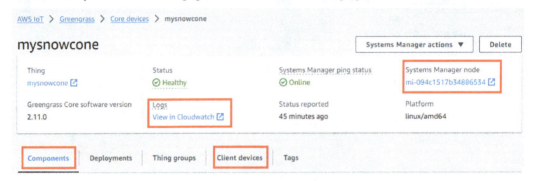

Figure 12.40 – Detail of the Snowcone device as a Greengrass Core device

Let's review what each of the elements highlighted in red represents:

- **Components** – This tab will show you which Greengrass components are installed and running on your device, as well as troubleshooting information if any are failing. In the case of this deployment, we have deployed and configured 19 Greengrass components. To see more details about them and their configuration, navigate to the **Deployments** tab and click on the **Deployment ID** link. In the upper right is an **Actions** dropdown. From that, select **Download as JSON**. An example of this is shown in the following figure:

Figure 12.41 – Downloading the component configuration as a JSON file

- **Logs** – One of the components we deployed was the Amazon CloudWatch agent. Click this link to see detailed log information coming in from the various components deployed in Greengrass as well as general operating system information about the EC2 instance:

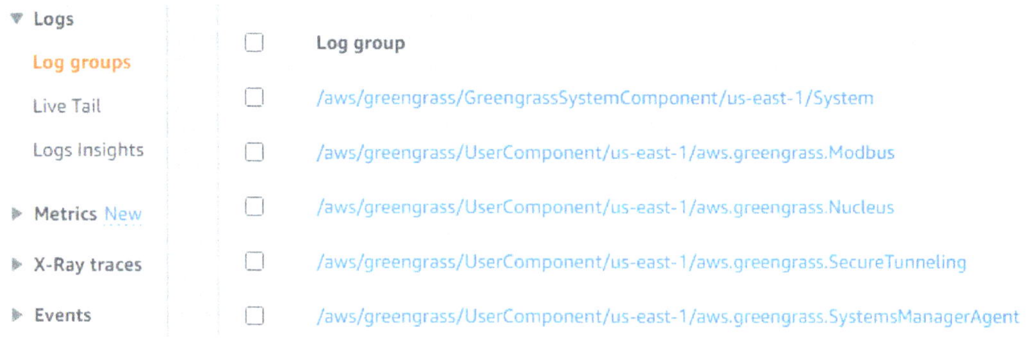

	Log group
▼ Logs	
Log groups	
Live Tail	/aws/greengrass/GreengrassSystemComponent/us-east-1/System
Logs Insights	/aws/greengrass/UserComponent/us-east-1/aws.greengrass.Modbus
▶ Metrics New	/aws/greengrass/UserComponent/us-east-1/aws.greengrass.Nucleus
▶ X-Ray traces	/aws/greengrass/UserComponent/us-east-1/aws.greengrass.SecureTunneling
▶ Events	/aws/greengrass/UserComponent/us-east-1/aws.greengrass.SystemsManagerAgent

Figure 12.42 – Viewing log groups in Amazon CloudWatch

- **Systems Manager** node – Another component included in the deployment was the Systems Manager agent. Click this link to view configuration information or perform tasks against the EC2 instance the same way you would any other:

▼ Tools	🗀 greengrass	Thu, 26 Oct 2023 13:48:52 GMT	root	root	drwxr-xr-x
File system	🗀 home	Thu, 26 Oct 2023 13:49:26 GMT	root	root	drwxr-xr-x
Performance counters	🗀 lib	Thu, 26 Oct 2023 13:58:09 GMT	root	root	dr-xr-xr-x
Processes	🗀 lib64	Thu, 26 Oct 2023 13:55:18 GMT	root	root	dr-xr-xr-x
Users and groups					
Execute run command	🗀 local	Fri, 19 Feb 2021 21:42:16 GMT	root	root	drwxr-xr-x
↗	🗀 media	Tue, 09 Apr 2019 19:57:27 GMT	root	root	drwxr-xr-x

Figure 12.43 – Viewing the filesystem inventory from AWS Systems Manager

- **Client Devices** – AWS IoT Greengrass v2 Core devices can be standalone endpoints if you wish. However, most customers use them as gateways for other IoT devices. This is especially true when using Greengrass on an AWS Snowcone device. The **Client Devices** tab is where you would initiate the cloud discovery procedure. This process will help you connect IoT things into Greengrass via this AWS Snowcone device, instead of them needing direct connectivity into AWS over the internet.

This is facilitated by the **MQTT Broker**, **MQTT Bridge**, **Client Auth**, and **IP Discovery** components that were included as part of this deployment.

Summary

In this chapter, we ran through a hands-on exercise that deployed a set of **Infrastructure As Code (IaC)** templates using AWS CloudFormation. This allowed you to deploy a complex and interrelated constellation of infrastructure components in minutes – something that could have taken hours to configure by hand in the AWS Management Console.

Next, we showed you how to install the AWS Snow CLI and configure AWS CLI v2 for use with the endpoints on an AWS Snow Family device. You also learned how to use these tools to unlock the device, create VNIs, and launch EC2 instances. We then covered how to configure the AWS IoT Greengrass v2 agent within an EC2 instance running on the AWS Snow Family device.

Finally, we did a walkthrough of how to connect to the EC2 instance remotely, and how to use the AWS IoT Greengrass v2 Core device as a gateway for sensors and other AWS IoT Core devices.

In the next chapter, you will walk through another hands-on exercise that uses IaC templates written in Terraform to deploy a distributed edge application that spans AWS Local Zones, AWS Wavelength Zones, and an AWS Region.

13

Deploying a Distributed Edge Computing Application

In this chapter, we will be using Terraform to deploy a distributed edge application on EKS that has a presence in three locations: an AWS Wavelength Zone, an AWS Local Zone, and the associated AWS Parent Region. The Wavelength and Local Zones are both in the same major metropolitan area. Depending upon the city you select, the Parent Region may be hundreds of miles away.

You will be able to see how a distributed EKS cluster that spans these edge locations can be put together, and clearly see the impact your entry point has on latency.

Check out the following video to view the Code in Action: `https://bit.ly/47LxSqd`

We'll cover the following topics:

- Deploying a distributed EKS cluster with Terraform
- Using Terraform to push a distributed application
- Testing the application from multiple sources
- Cleanup
- Notes

Deploying a distributed EKS cluster with Terraform

In the last chapter, we used **Infrastructure As Code (IaC)** written in AWS CloudFormation. In this chapter, we will use a different tool – HashiCorp Terraform. It is similar to AWS CloudFormation in that you create template files that describe interrelated infrastructure components you want created in your AWS account.

However, Terraform supports more than just AWS via a lengthy `list of providers`. This makes it popular with those who need to manage resources in multiple clouds – as well as on-premise elements such as VMware virtual machines or even bare-metal servers. AWS is a strong supporter of Terraform, with over 2 billion downloads of the `AWS Terraform provider` since 2014.

The IaC provided supports the following 15 US cities: Atlanta, Boston, Chicago, Dallas, Denver, Houston, Las Vegas, Los Angeles, Miami, Minneapolis, New York City, Phoenix, San Francisco, Seattle, and Washington DC.

It also supports the following four non-US cities: London, Osaka, Seoul, and Tokyo.

> **Note about AWS Local Zone availability**
>
> Some major cities that have AWS Wavelength Zones do not also have AWS Local Zones because there is already a full AWS Region there. For these areas, instead of a Local Zone, one of the normal Availability Zones in that region will be used.
>
> The cities affected by this are London, Osaka, San Francisco, Seoul, Tokyo, and Washington DC.

Architecture

The next figure gives a high-level overview of what the Terraform IaC in this exercise will deploy when the selected edge city is Atlanta:

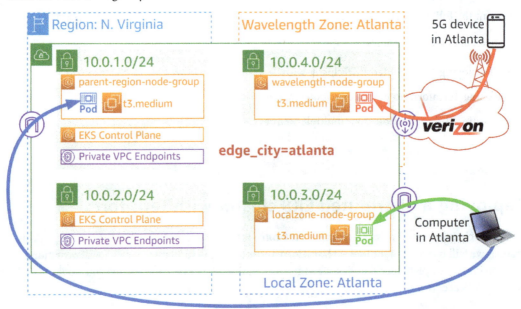

Figure 13.1 – Distributed edge application overview when edge_city = atlanta

The purposes of this design and associated Terraform IaC are as follows:

- Showing you the difference in user experience for three different ways of accessing the same application

- Familiarizing you with using Terraform to deploy EKS in a distributed fashion

- Giving you an example of how to manage a Kubernetes deployment for a sample distributed application

> **Note about your physical location**
>
> You should deploy this IaC to whichever of the 20 supported `edge_city` options you are geographically present in. Obviously, if you do not live in one of these cities, you will need to choose the one closest to you. However, you should be aware that the difference in latency between the three entry points becomes less and less dramatic the farther you physically are from the edge city you deploy to.

Components

The Terraform IaC for this stage will deploy the following components into your AWS account:

- **EKS cluster** – The control plane elements will be deployed into the parent region.

- **VPC** – This includes the associated subnets, security groups, and routing tables needed to support operation in all three zone types we will be using.

- **Private endpoints** – 11 VPC endpoints will be deployed into the parent region. These allow the EKS worker nodes to communicate with services such as EKS, ECR, S3, SSM, and CloudWatch without needing to send that traffic out over the internet. This is particularly important with AWS Wavelength Zones.

- **Self-managed EKS node groups** – As you'll recall from earlier chapters, EKS managed node groups are not supported in AWS Wavelength. Therefore, the IaC will deploy three self-managed node groups, each of which will contain a single `t3.medium` instance.

Cost

Understandably, cost is always top of mind when deploying resources into your AWS account.

To give you a sense of the costs incurred while doing the exercises in this chapter, please reference the next figure. It is a snapshot from AWS Cost Explorer of an account that ran all of these components for 24 hours in the eu-west-2 (London) region:

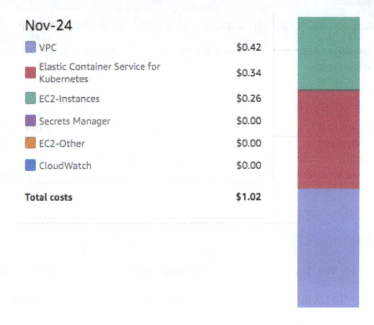

Figure 13.2 – 24 hours of charges in the eu-west-2 (London) region

Keep in mind that costs for things such as EC2 instances vary slightly between regions, but this should give you a ballpark understanding.

Setting up your environment

In this section, we will walk through configuring a workstation that you will use to download the repository from GitHub and deploy the IaC components inside.

Workstation selection

The first task you must perform is to set up the environment of whatever computer you wish to use to deploy and manage this infrastructure from. All of the example commands and figures in this chapter assume you are using an Ubuntu Linux 22.x environment.

It is, therefore, recommended that you use one of the following configurations:

- **Windows 10/11 (WSL)** – If your workstation is Windows-based, it is strongly recommended that you install Windows Subsystem for Linux (WSL), specifically the Ubuntu LTS 22(x) environment available from the Windows Store:

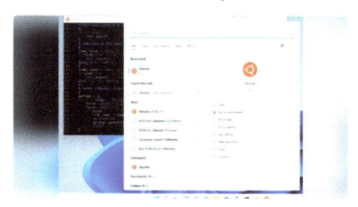

Figure 13.3 – Ubuntu 22.04.2 LTS on the Windows Store

This is the quickest and easiest way to get your environment set up and working in Windows. While it is possible to perform all of the tasks in this chapter with native Windows commands in PowerShell, the specifics of doing so are not covered here.

- **Native Linux** – If you prefer to use a cloud-based Linux workstation in your AWS account, an EC2 instance running Ubuntu 22(x) (or any Debian-based distribution) is ideal. Ensure it has internet access via a public IP or **network address translation** (**NAT**) gateway. If you have never set up a Linux-based bastion host in AWS, this Quick Start is recommended: `https://aws.amazon.com/solutions/implementations/linux-bastion/`.

- **AWS CloudShell** is a serverless option that works fine; just be aware you may have to reconfigure your environment if you walk away halfway through the exercises and return the next day. `AWS Cloud9` is a fully functional IDE that is the most feature-rich out of all three options.

Installing and configuring AWS CLI v2

If you do not have AWS CLI v1 installed at all, simply execute the following commands:

```
curl \
 "https://awscli.amazonaws.com/awscli-exe-linux-x86_64.zip" \
 -o "awscliv2.zip"
unzip awscliv2.zip
sudo ./aws/install
```

If you need to upgrade from AWS CLI v1 to v2, use the following commands instead:

```
curl \
 "https://awscli.amazonaws.com/awscli-exe-linux-x86_64.zip" \
 -o "awscliv2.zip"
unzip awscliv2.zip
sudo ./aws/install \
 --bin-dir /usr/local/bin \
 --install-dir /usr/local/aws-cli \
 --update
```

Whichever method you use, the installation process can take as long as 5 minutes. See User Guide for Version 2 in the *AWS Command Line Interface* documentation for more details if you run into any trouble.

Next, configure the AWS CLI with administrative credentials for your account, and set the default region. The access key and secret access key can be obtained from the **AWS Management Console** under **IAM** > **Users** > **Credentials**:

```
aws configure \
 set aws_access_key_id "your_access_key"

aws configure \
 set aws_secret_access_key "your_secret_key"

aws configure \
 set region "your_region"
```

To make sure you are connected to your AWS account, execute the following command. It should show your account ID. The next figure gives an example of what you should see:

Figure 13.4 – Validating the account ID AWS CLI is connected to

Cloning the code repository from GitHub

Clone the repository from GitHub with the following command:

```
git clone \
  https://github.com/PacktPublishing/Edge-Computing-with-Amazon-Web-
  Services.git
```

The following figure gives an example of what you should see:

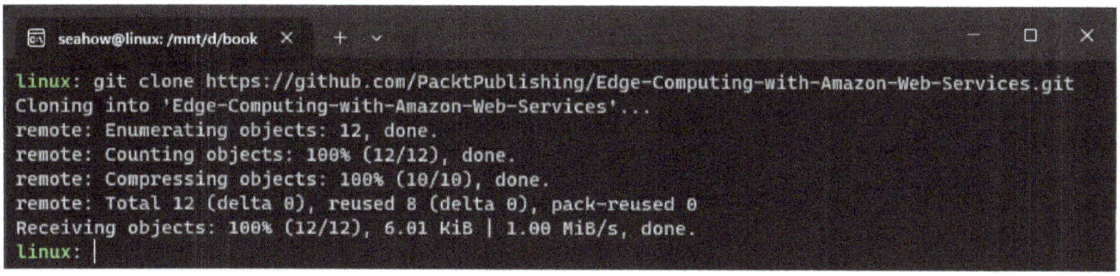

Figure 13.5 – Cloning the repository from GitHub

Installing kubectl v1.28

Kubernetes provides `kubectl`, a command-line tool that allows you to perform commands against a cluster's control plane via the Kubernetes API. Think of it as the same type of thing as the AWS CLI, only for specific operations within Kubernetes itself.

The following commands will install `kubectl` version 1.28. It is recommended you install this specific version, as the EKS cluster will match:

```
sudo apt-get install -y apt-transport-https
sudo apt-get install -y ca-certificates curl

curl -fsSL https://pkgs.k8s.io/core:/stable:/v1.28/deb/Release.key |
```

```
sudo gpg --dearmor -o /etc/apt/keyrings/kubernetes-apt-keyring.gpg

echo 'deb [signed-by=/etc/apt/keyrings/kubernetes-apt-keyring.gpg]
https://pkgs.k8s.io/core:/stable:/v1.28/deb/ /' | sudo tee /etc/apt/
sources.list.d/kubernetes.list

sudo apt-get update
sudo apt-get install -y kubectl
```

For more details, see `https://kubernetes.io/docs/tasks/tools/install-kubectl-linux/`.

Installing Terraform

Unlike AWS CloudFormation, which executes inside AWS, Terraform will be doing all of its work and maintaining all of its state information directly on your workstation. Note that it is possible to use a hosted version of Terraform, and it is also possible to configure your state management such that it resides on S3 or in RDS – but such things are out of scope for this chapter.

The following commands will add the HashiCorp repository to your `apt` configuration, which will then allow you to install Terraform via `apt`:

```
wget -O- https://apt.releases.hashicorp.com/gpg | sudo gpg --dearmor
-o /usr/share/keyrings/hashicorp-archive-keyring.gpg

echo "deb [signed-by=/usr/share/keyrings/hashicorp-archive-keyring.
gpg] https://apt.releases.hashicorp.com $(lsb_release -cs) main" |
sudo tee /etc/apt/sources.list.d/hashicorp.list

sudo apt update

sudo apt install terraform -y
```

You are now ready to begin deploying the IaC packages into your AWS account.

Deploying the EKS cluster

In this section, you will initialize and deploy the Terraform IaC in the `distributed-application/eks-cluster` directory of the repository you downloaded.

NOTE: Replace `<your_edge_city>` with one of the following values:

atlanta	denver	miami	seattle	seoul
boston	houston	minneapolis	london	tokyo
chicago	lasvegas	newyorkcity	osaka	washingtondc
dallas	losangeles	phoenix	sanfrancisco	

Step 1 – Entering the distributed-application/eks-cluster directory

```
cd distributed-application/eks-cluster
```

Step 2 – Initializing Terraform providers/modules in this directory

The `terraform init` command will download and configure the specific provider versions and modules specified in the IaC in this directory:

```
terraform init
```

Here's an example of what you should see:

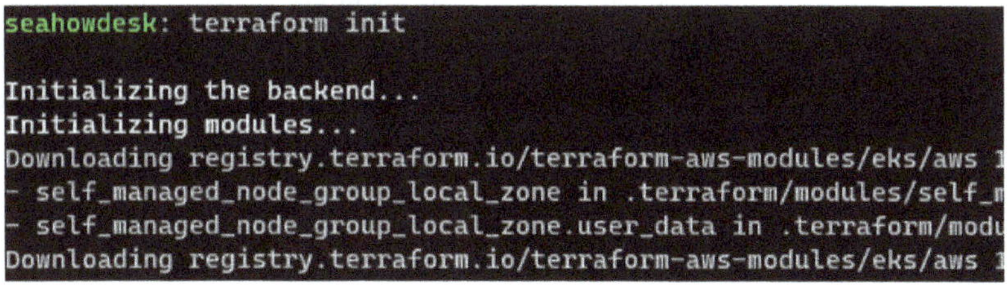

Figure 13.6 – Initializing Terraform in the eks-cluster directory

Step 3 – Starting the deployment with Terraform

Use the `terraform apply` command as follows:

```
terraform apply -var=edge_city=<your_edge_city> -auto-approve
```

Here's an example of what you should see:

```
seahowdesk: terraform apply -var=edge_city=atlanta -auto-approve
data.aws_eks_cluster_auth.default: Reading...
module.self_managed_node_group_local_zone.data.aws_ami.eks_default[0]: Reading...
data.aws_caller_identity.current: Reading...
module.self_managed_node_group_parent_region.data.aws_partition.current: Reading...
module.self_managed_node_group_wavelength.data.aws_caller_identity.current: Reading...
module.self_managed_node_group_parent_region.data.aws_caller_identity.current: Reading...
data.aws_partition.current: Reading...
```

Figure 13.7 – Applying Terraform IaC for eks-cluster

> **Note**
>
> This will take around 10 minutes to complete.

Step 4 – Using the AWS CLI to configure kubectl

This will be needed to validate your EKS deployment. Issue the following command:

```
aws eks update-kubeconfig --name distributedcluster --region <your_
region>
```

Replace `<your_region>` with the AWS parent region you are targeting, as in the following example:

```
seahowdesk: aws eks update-kubeconfig --name distributedcluster --region us-east-1
Added new context arn:aws:eks:us-east-1:          :cluster/distributedcluster to
```

Figure 13.8 – Using the AWS CLI to update configuration for kubectl

Step 5 – Using kubectl to verify your EKS deployment

Execute the following `kubectl` command to see the status of your EKS cluster:

```
kubectl get nodes
```

> **NOTE:**
>
> The node on the `10.0.4.0/24` subnet is located in AWS Wavelength, and it is normal for it to show a status of `NotReady`. This will be addressed in the next section.

Here's an example of what you should see:

```
seahowdesk: kubectl get nodes
NAME                       STATUS     ROLES     AGE      VERSION
ip-10-0-1-248.ec2.internal    Ready      <none>    3m48s    v1.28.3-eks-e71965b
ip-10-0-3-93.ec2.internal     Ready      <none>    2m55s    v1.28.3-eks-e71965b
ip-10-0-4-176.ec2.internal    NotReady   <none>    8s       v1.28.3-eks-e71965b
```

Figure 13.9 – Retrieving the status of the EKS nodes using kubectl

Step 6 – Changing back to the root directory

In preparation for the next section, return to the root of the `distributed-application` directory within the repository you downloaded:

```
cd ..
```

Using Terraform to push a distributed application

Now that you have EKS deployed, we will use a separate set of Terraform IaC to deploy the distributed application itself.

Components

- **Kubernetes deployments** – Three deployments in separate namespaces and with separate node taints. This will force pods to deploy to either the parent region, the Local Zone, or the Wavelength Zone, as appropriate.

- **Kubernetes services** – A simple `NodePort` service will allow traffic from each node's public IP address into those locations. This will allow you to test accessing containers in EKS via three different paths into AWS.

Deploying the distributed application

In this section, you will initialize and deploy the Terraform IaC in the `distributed-application/kubernetes-config` directory of the repository you downloaded.

> **NOTE:**
> You must replace `<your_edge_city>` with the same value you used in the previous section when you deployed the EKS cluster.

Step 1 – Entering the kubernetes-config directory

```
cd kubernetes-config
```

Step 2 – Initializing Terraform providers/modules in this directory

The `terraform init` command will download and configure the specific provider versions and modules specified in the IaC in this directory:

```
terraform init
```

Step 3 – Starting the deployment with Terraform

Use the `terraform apply` command as follows:

```
terraform apply -var=edge_city=<your_edge_city> -auto-approve
```

> NOTE:
>
> Remember to use the same value for `<your_edge_city>` that you used previously.

Here's an example of what you should see:

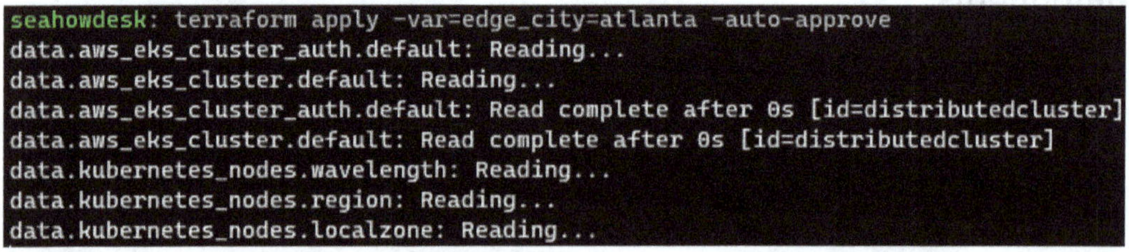

```
seahowdesk: terraform apply -var=edge_city=atlanta -auto-approve
data.aws_eks_cluster_auth.default: Reading...
data.aws_eks_cluster.default: Reading...
data.aws_eks_cluster_auth.default: Read complete after 0s [id=distributedcluster]
data.aws_eks_cluster.default: Read complete after 0s [id=distributedcluster]
data.kubernetes_nodes.wavelength: Reading...
data.kubernetes_nodes.region: Reading...
data.kubernetes_nodes.localzone: Reading...
```

Figure 13.10 – Applying Terraform IaC for kubernetes-config

Step 4 – Copying the output URLs

When the previous `terraform apply` command is finished, you should see an output that looks like this:

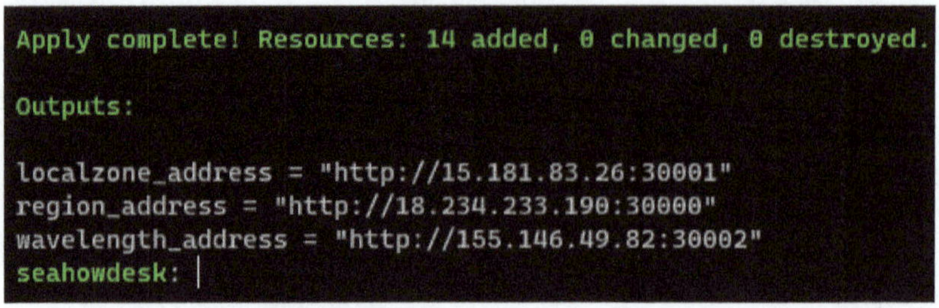

```
Apply complete! Resources: 14 added, 0 changed, 0 destroyed.

Outputs:

localzone_address = "http://15.181.83.26:30001"
region_address = "http://18.234.233.190:30000"
wavelength_address = "http://155.146.49.82:30002"
seahowdesk: |
```

Figure 13.11 – URL output by terraform apply in kubernetes-config

Save these values. They will be needed in the next section.

Testing the application from multiple sources

Assuming the Terraform IaC you applied in the previous sections deployed without any problems, you will now be able to connect to the test application from three different entry points. One entry point is in AWS Wavelength, one is in an AWS Local Zone, and one is in the parent AWS Region.

In this section, you will connect to all three entry points using a desktop/laptop on your home Wi-Fi or similar standard internet connection. Then, you will connect to the same three URLs from a mobile device that is on the appropriate carrier network for the AWS Wavelength Zone in question.

Checking the user experience from your desktop/laptop

From a desktop connected to the internet, ping all three nodes, and note the RTT in milliseconds:

```
ping 15.181.83.26 -c 2
ping 18.234.233.190 -c 2
ping 155.146.49.82 -c 2
```

Here's an example of what you should see:

```
seahowdesk: ping 15.181.83.26 -c 2
PING 15.181.83.26 (15.181.83.26) 56(84) bytes of data.
64 bytes from 15.181.83.26: icmp_seq=1 ttl=239 time=108 ms
64 bytes from 15.181.83.26: icmp_seq=2 ttl=239 time=108 ms

--- 15.181.83.26 ping statistics ---
2 packets transmitted, 2 received, 0% packet loss, time 1001ms
rtt min/avg/max/mdev = 107.617/108.020/108.423/0.403 ms
```

Figure 13.12 – Pinging the IP of the entry point in the AWS parent region

Now, visit all three URLs in a web browser, again from your desktop/laptop on a standard internet connection. You should see the following page:

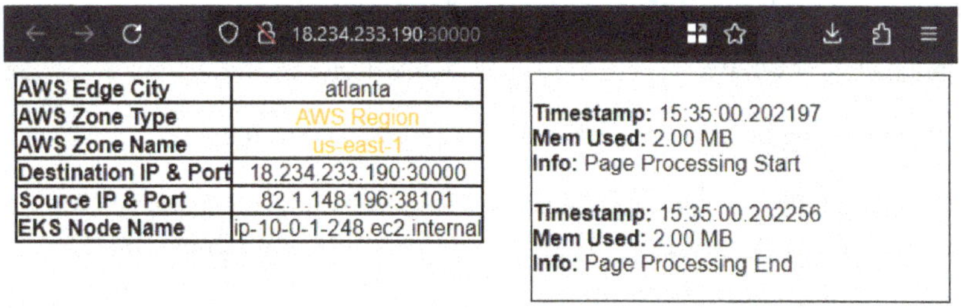

Figure 13.13 – Viewing the entry point in the AWS parent region from a desktop/laptop

The page will show some information about where you have connected to and where your connection originated:

AWS Edge City	atlanta
AWS Zone Type	AWS Local Zone
AWS Zone Name	us-east-1-atl-1a
Destination IP & Port	15.181.83.26:30001
Source IP & Port	82.1.148.196:38097
EKS Node Name	ip-10-0-3-93.ec2.internal

Timestamp: 15:34:58.072439
Mem Used: 2.00 MB
Info: Page Processing Start

Timestamp: 15:34:58.072487
Mem Used: 2.00 MB
Info: Page Processing End

Figure 13.14 – Viewing the entry point in the AWS Local Zone from a desktop/laptop

However, when you attempt to visit the URL for the AWS Wavelength Zone, the connection will time out:

Figure 13.15 – Timeout when connecting to AWS Wavelength from a desktop/laptop

Checking the user experience from your mobile device

Now, let's test the benefit of AWS Wavelength from your mobile device on the appropriate carrier network.

As a reminder, the mapping of carriers to cities the IaC supports is as follows:

- **Verizon** – Atlanta, Boston, Chicago, Dallas, Denver, Houston, Las Vegas, Los Angeles, Miami, Minneapolis, New York City, Phoenix, Seattle, San Francisco, Washington DC
- **Vodafone** – London

- **KDDI** – Tokyo, Osaka

- **SK Telecom** – Seoul

> **Note**
>
> If you do not have a mobile device on the appropriate carrier network, you will not be able to do the exercises in this section.

First, let's visit the URL for the AWS Wavelength Zone in a browser on your mobile device:

AWS Edge City	london
AWS Zone Type	AWS Wavelength
AWS Zone Name	eu-west-2-wl1-lon-wlz-1
Destination IP & Port	141.195.195.159:30002
Source IP & Port	148.252.140.81:49712
EKS Node Name	ip-10-0-4-60.eu-west-2.compute.internal

Figure 13.16 – Viewing the entry point in the AWS Wavelength Zone from a mobile device

Unlike the timeout you received when accessing this URL from your desktop/laptop, you should see an output similar to what is shown in the preceding figure.

> **Note**
>
> If you have trouble connecting, and you are on the correct carrier network, try restarting the pod in the AWS Wavelength Zone by deleting it with `kubectl`. This will force Kubernetes to recreate the pod, which has the side effect of restarting it.

First, use `kubectl` to get a list of pod names in the `distributed-app-wavelength` namespace. There should only be one:

```
kubectl get pods -n distributed-app-wavelength
```

The output should look like that shown in the following figure:

```
seahowdesk: kubectl get pods -n distributed-app-wavelength
NAME                                        READY   STATUS
wavelength-deployment-54965dddcf-c2fwv      1/1     Running
```

Figure 13.17 – Using kubectl to get the pod name in AWS Wavelength

Now, use the pod name in the following command to delete/recreate it:

```
kubectl delete pod wavelength-deployment-54965dddcf-c2fwv -n
distributed-app-wavelength
```

Kubernetes will recreate the pod within seconds of this action. Try again from your mobile device's browser to access the URL.

Next, let's check the RTT to all three IP addresses from your mobile device:

Figure 13.18 – Pinging the AWS parent region from a mobile device

ManageEngine Ping Tool by Zoho

Most mobile devices don't come with a native `ping` utility. Therefore, you will need to install one. One approach is to install ManageEngine Ping Tool by Zoho. It is free and available for both IOS and Android.

After you've pinged all three IP addresses and compared their RTTs, let's use another function of ManageEngine Ping Tool to view average response times:

Figure 13.19 – Measuring average response time

If you are connected via 5G to the appropriate carrier network and are geographically located in the edge city where you deployed the IaC, the average response time to the AWS Wavelength Zone should be dramatically lower than the other two.

Cleanup

To remove all of the components deployed by the Terraform IaC, perform the following steps. Note that removal happens in reverse order of deployment.

Step 1 – Using terraform destroy to remove kubernetes-config elements

Since you should still be in the `kubernetes-config` directory, you can proceed immediately to the `terraform destroy` command shown here:

```
terraform destroy -var=edge_city=atlanta -auto-approve
```

When the process is complete, you should see an output similar to that shown in the following figure:

```
kubernetes_node_taint.wavelength[0]: Destruction complete after 0s
kubernetes_node_taint.region[0]: Destruction complete after 0s
kubernetes_node_taint.localzone[0]: Destruction complete after 0s
aws_eip.wavelength_eip: Destruction complete after 2s
kubernetes_namespace.distributed-app-localzone: Destruction complete after 8s
kubernetes_namespace.distributed-app-wavelength: Destruction complete after 8s
kubernetes_namespace.distributed-app-region: Destruction complete after 8s

Destroy complete! Resources: 14 destroyed.
seahowdesk:
```

Figure 13.20 – Using terraform destroy to remove kubernetes-config elements

Step 2 – Using terraform destroy to remove eks-cluster elements

First, you will need to change directory to eks-cluster using the following command:

```
cd ../eks-cluster
```

Once you're there, you can issue the exact same terraform destroy command you did in step 1:

```
terraform destroy -var=edge_city=atlanta -auto-approve
```

Upon successful completion, you should see an output similar to that shown in the following figure:

```
aws_iam_role_policy_attachment.k8s-distributed-AmazonEKSClusterPolicy: Destroying.
aws_subnet.parent-region-subnet-a: Destroying... [id=subnet-037d7c5a7b10d0c5c]
aws_iam_role_policy_attachment.k8s-distributed-AmazonEKSVPCResourceController: Des
aws_iam_role_policy_attachment.k8s-distributed-AmazonEKSClusterPolicy: Destruction
aws_iam_role.k8s-distributed-cluster: Destroying... [id=distributedcluster]
aws_iam_role.k8s-distributed-cluster: Destruction complete after 1s
aws_subnet.parent-region-subnet-a: Destruction complete after 1s
aws_subnet.parent-region-subnet-b: Destruction complete after 1s
aws_vpc.k8s-distributed: Destroying... [id=vpc-01d51e5a01b78d4c9]
aws_vpc.k8s-distributed: Destruction complete after 1s

Destroy complete! Resources: 61 destroyed.
seahowdesk:
```

Figure 13.21 – Using terraform destroy to remove eks-cluster elements

All AWS components of the Terraform IaC stack deployed are now deleted, and no charges will accrue going forward.

Notes

If you run into trouble or are simply curious to explore more regarding your EKS cluster, please be aware of the following:

- **Systems Manager** – The SSM agent is installed and IAM policies/roles are configured such that you can connect to your EKS worker nodes via Session Manager, which is a feature of Systems Manager. If you need to connect to them directly for troubleshooting, this is a quick and easy way to do so.

- **CloudWatch** – If you want to troubleshoot anything about the EKS cluster, you should enable CloudWatch logging. Navigate to **EKS** > **Clusters** > **distributedcluster**, then click the **Observability** tab. In the section marked **Control plane logging**, click **Manage logging**. Enable all of the slider buttons and save your changes. Now, you can view extensive Kubernetes logs in **CloudWatch** > **Logs** > **Log Groups** > **/aws/eks/distributedcluster/cluster**.

Summary

In this chapter, we ran through a hands-on exercise that deployed a set of IaC templates using Terraform. This allowed you to deploy a complex and interrelated constellation of infrastructure components in minutes – something that could have taken hours to configure by hand in the AWS Management Console.

Next, we showed you how to use another set of IaC templates to deploy a containerized distributed application that spanned an AWS Region, an AWS Local Zone, and an AWS Wavelength Zone. You can now use both IaC stacks as the basis of further development of your own distributed applications.

Finally, we gave you some pointers on how to evaluate the impact on user experience the physical location of the application entry point has.

In the next, and final, chapter, we will review the future of AWS edge computing services.

14
Preparing for the Future of Edge Computing with AWS

In this, our final chapter, we will cover some of the key business drivers behind the explosive growth in edge computing to help you plan for future developments in the industry. We will also review how to plan for growth in your architecture so that what you put in place now will continue to function if your application's demand scales.

Finally, we'll review some best practices and pitfalls to avoid with any edge computing service so that you'll be prepared no matter what new announcements AWS makes.

Here are the main topics covered in this chapter:

- Business drivers for growth in edge computing
- Building a solid foundation for the future
- Patterns and anti-patterns

Business drivers for growth in edge computing

There are two main categories of emerging business needs that are predicted to drive explosive growth in new edge computing applications, as well as rapid and unpredictable scaling of those applications once they are in place. The first category is those drivers that fall under the Industry 4.0 zeitgeist, and the second consists of those necessitated by data sovereignty and privacy concerns.

Industry 4.0

Industry 4.0 is a term that was popularized by the World Economic Forum founder and executive chairman Klas Schwab in his 2017 book, *The Fourth Industrial Revolution*. It predicts a number of fundamental shifts in how production and supply networks function around the world. These are expected to be driven by imminent breakthroughs in emerging technologies that fuse digital, physical, and human components.

The shift to Industry 4.0 is not a mere technological upgrade but a strategic move to address a range of business challenges and opportunities. From expected gains in operational efficiency to increased ability to meet the demands of a rapidly evolving market, Industry 4.0 practices are now front and center in the technology strategies of governments and major manufacturers worldwide.

Given this background, let's review some of the expected benefits of any edge computing solution to guide your architecture. These may not be stated upfront as requirements, but you should at least plan for the possibility that your architecture will be asked to deliver on them at some point in the future:

- **Increased efficiency and productivity**

  One of the primary motivations for adopting Industry 4.0 practices is the significant gain in operational efficiency. Automation, digitalization, and smart technologies allow for faster, more efficient production processes, reducing manual labor and minimizing errors. Smart factories can optimize resource utilization, streamline supply chains, and improve overall productivity.

- **Enhanced data analysis and decision-making**

  Industry 4.0 introduces advanced data analytics capabilities. By leveraging big data, IoT devices, and real-time analytics, businesses can gain deeper insights into their operations. This data-driven approach enables more informed decision-making, predictive maintenance, and proactive problem-solving.

- **Customization and flexibility**

  Today's market demands high levels of customization and flexibility in manufacturing. Industry 4.0's smart systems allow for agile production lines that can be quickly adapted to change orders or customize products according to individual customer demands.

- **Improved product quality**

  Advanced monitoring and **Quality Control** (**QC**) enabled by smart technologies ensure higher standards of product quality. Automated quality checks and real-time feedback loops help maintain consistent product quality and reduce waste due to defects.

- **Competitive advantage**

  Implementing Industry 4.0 practices can provide a significant competitive advantage. It not only enhances operational efficiency and product quality but also enables businesses to respond swiftly to market changes and customer demands, positioning them ahead in the competitive landscape.

- **Sustainability and environmental concerns**

  Industry 4.0 aligns with the growing emphasis on sustainability. Efficient resource utilization, reduced waste, and energy-efficient operations contribute to more environmentally friendly manufacturing processes.

- **Customer expectations and market demand**

  With the digital transformation of the economy, customers expect faster service, higher-quality products, and more customization options. Industry 4.0 enables businesses to meet these evolving customer expectations.

- **Globalization and supply chain complexity**

  As businesses operate in an increasingly globalized market, managing complex supply chains becomes crucial. Industry 4.0 practices offer better visibility and control over global supply chains, improving logistics and distribution.

- **Workforce development and skills enhancement**

  The shift toward smart technologies necessitates a more skilled workforce. Adopting Industry 4.0 practices drives the need for training and development, leading to a more skilled and efficient workforce.

- **Risk management and compliance**

  Advanced monitoring and predictive analytics help in better risk management. Also, Industry 4.0 can aid in ensuring compliance with regulatory standards and safety protocols.

Data sovereignty and privacy concerns

You are likely aware that, in terms of privacy, the world today is not the same as it was even 15 years ago. For example, as of 2022, there were a billion digital cameras operating worldwide. Even private residences now commonly record video of the street they are on, never mind the mobile devices that are quickly activated whenever something even mildly interesting occurs.

Governments worldwide are responding with new regulatory frameworks that seek to enforce their privacy laws amid this situation. They seek to answer questions such as these:

- How can they make sure that someone's face, personal information, and unproven allegations aren't broadcast to billions around the world?

- How can they ensure sensitive information about their country's infrastructure operations doesn't fall into the hands of groups that might want to do their citizens harm?

- How do they give their citizenry a say in what corporations do with the ever-sprawling datasets about their personal life?

- How can they even capture an up-to-date snapshot of these datasets, and what is being done with them?

With that as context, let's run through some specific things to keep in mind when architecting an edge computing solution.

Regulatory compliance

With the advent of stringent data protection laws such as the **General Data Protection Regulation (GDPR)** in Europe, the **California Consumer Privacy Act (CCPA)** in the US, and similar regulations worldwide, businesses are under pressure to manage data responsibly. Industry 4.0 technologies, such as advanced data analytics and IoT, must be designed and operated in compliance with these laws. This involves ensuring data is collected, processed, and stored in ways that respect user privacy and consent.

Data sovereignty

This concept involves adhering to laws and regulations about the storage and handling of data within the geographical borders of a country. Industry 4.0 practices have to account for data sovereignty, especially for multinational companies that operate across different jurisdictions with varying data laws.

Secure data handling

As Industry 4.0 systems collect and process large volumes of data, including potentially sensitive information, ensuring the security of this data is paramount. This involves implementing robust cybersecurity measures, secure data storage solutions, and secure communication channels to prevent data breaches and unauthorized access.

Privacy by design

This approach involves integrating data privacy and security features right from the initial design phase of products and systems. In the context of Industry 4.0, this means designing IoT devices, cloud services, and data analytics platforms with built-in privacy controls and security protocols.

Impact on the supply chain

Data privacy regulations also impact the supply chain, as companies need to ensure their partners and suppliers comply with relevant data laws. This requires a thorough vetting process and, often, the implementation of standardized data handling practices across the supply chain.

Training and awareness

Businesses must also invest in training their workforce to be aware of data privacy and regulatory requirements. This is essential to ensure that employees handling data are aware of compliance standards and the importance of protecting consumer information.

Data transparency and consumer trust

There's a growing demand from consumers for transparency in how their data is used. Companies embracing Industry 4.0 need to not only comply with laws but also build trust by being transparent about their data practices. This includes clear communication about data usage and ensuring customers' rights regarding their data are respected.

Regular audits and updates

Compliance with data privacy laws is not a one-time task but requires ongoing effort. Regular audits, assessments, and updates to policies and technologies are necessary to ensure continued compliance, especially as laws and regulations evolve.

Now that we've reviewed key business drivers for growth in demand for edge computing applications, let's shift to how you can go about building them with future changes in mind.

Building a solid foundation for the future

Planning for the future in the context of edge computing on AWS involves anticipating and adapting to new technologies, scaling strategies, and changing market needs. This section covers some of the ways you can go about this with the extended set of AWS services and features we haven't covered yet.

Taking a systematic approach

The best way to ensure your application is ready for the future is to build an appropriate architectural and operational foundation from the start. There are two key elements to this – establishing policies and practices and translating those into specific blueprints or configuration patterns.

AWS Cloud Foundation

The `Establishing Your Cloud Foundation on AWS` section of the `AWS Architecture Center` is an excellent starting point for doing this. It contains guidance on how to establish foundational practices as part of a `capability-based approach` for the applications you build on AWS. This can be thought of as a stripped-down, AWS-specific version of something such as the **Information Technology Infrastructure Library (ITIL)**.

Examples of such capabilities you should include in your architecture are discussed next.

Identity Management and Access Control

The `Identity Management and Access Control (IMAC) capability` guides you in implementing and monitoring your configuration of `AWS Identity and Access Management (IAM)` – something that is used in every single architecture built on AWS.

This may well represent the most difficult thing to get right the first time, and you will benefit greatly from the guidance found here. It covers things such as how to use `AWS IAM Identity Center` for **multi-factor authentication (MFA)** and federation with other identity sources such as Active Directory, Google Workspace, or Okta.

Workload isolation

The `Workload Isolation capability` includes best practices for `organizing your AWS environment` using a multi-account structure using `AWS Organizations and service control policies (SCPs)`.

Log storage

One of the most important things you must be able to do is quickly produce detailed reports on the flow of data through your application – especially who accessed what data and when. This might be to answer an auditor's questions or to cooperate with authorities when a breach is suspected.

The `Log Storage capability` involves using things such as `VPC Flow Logs` to monitor traffic at the network level, `AWS CloudTrail` to log access to the AWS API or Management Console, or AWS CloudWatch to centrally aggregate logs from within your applications.

Governance

The `Governance capability` helps you to define your organizational policies for business and regulatory compliance. This includes using services such as `AWS Artifact` to find published compliance reports for AWS' portion of the shared responsibility model. To assist with customer responsibilities, AWS publishes extensive `guidance on how to manage risk and compliance` for your applications running in the cloud.

Change management

The `Change Management capability` helps you manage risk and minimize the impact of planned changes to your production applications. This includes organizational processes such as sending planned changes that have an unknown risk profile to a **Change Advisory Board (CAB)** as **Requests for Change (RFCs)**. This, in turn, could be informed by using `AWS Config` as a **configuration management database (CMDB)** for your resources in the cloud.

Using the AWS Well-Architected Framework

In the same way the AWS Cloud Foundations can be thought of as a lightweight version of ITIL practices, the `AWS Well-Architected Framework` can be thought of as an AWS-specific take on enterprise architecture frameworks such as Zachman or TOGAF. It is organized into six pillars – operational excellence, security, reliability, performance, cost optimization, and sustainability.

The `AWS Well-Architected Tool` is available at no cost in the AWS Management Console. It is a guided walkthrough of many subcomponents of each of the six pillars. It provides specific architectural guidance and best practices in the context of specific AWS services and configurations. It helps you review the state of your applications to identify opportunities for improvement and track your progress toward them over time.

`AWS Trusted Advisor` is a service that will continuously evaluate your AWS environment against the best practices outlined in the AWS Well-Architected Tool. Note that even if you use this service, it is still important that you walk through the manual exercises in the AWS Well-Architected Tool to connect the intent of your organizational policies and goals with how things are configured.

Infrastructure as code

It is vital that you implement an operational policy whereby nothing gets changed within a production AWS account except by some form of **Infrastructure as Code (IaC)**. AWS CloudFormation or Terraform templates that are part of a CI/CD pipeline in AWS CodeBuild or GitLab are excellent starting points.

This is likely to be more work than simply logging in to the AWS Management Console and manually doing everything – at first, anyway. Once your team gets the hang of these tools, the visibility thus gained will yield exponential gains in the speed with which they can implement new features or respond to problems in the application.

Most importantly, you will be able to easily assess the impact of any proposed changes because you can deploy an exact duplicate of the application – including its underlying infrastructure – into a test account before pushing it to production.

Now that you're familiar with a systematic approach to building your edge computing applications, let's take a look at some recommendations as well as pitfalls to avoid.

Patterns and anti-patterns

This section will review recommended architectural patterns you should try to follow. It also touches on anti-patterns to outline some key pitfalls to avoid when architecting a distributed edge application on AWS.

Leveraging cloud-native services

It would certainly be possible to develop a globally distributed Industry 4.0 application without using any cloud-native services. Some customers are inclined to do this because they wish to reserve the option of moving their application to another cloud service provider or to an on-premise data center. The trouble with this approach is that you are forced to use only least-common-denominator features that are common across all CSPs or private clouds. In the context of AWS, this means you can only use standard S3 buckets, a limited set of EC2 instance types, and only the most basic features of supporting functions such as VPC. This has the effect of making your task of application design far more complex.

Simple IoT application example – the non-native version

Let's take the example of an application that ingests MQTT messages from IoT devices and hosts a dashboard that allows a user to browse the current state of these devices. This could absolutely be built using `Mosquitto` running on Ubuntu inside an EC2 instance. The dashboard could then be built on top of `Grafana` running on the same instance:

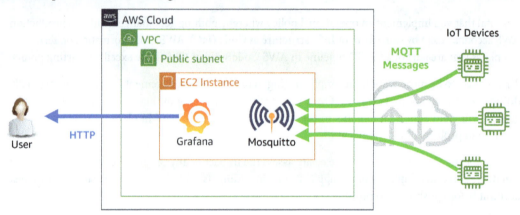

Figure 14.1 – Simple non-cloud-native IoT application

The preceding diagram shows this arrangement. From the perspective of cross-cloud portability, this is great. However, this design is not ideal when looked at from the following perspectives:

- **Resilience** – What if the EC2 instance crashes or its configuration is mistakenly deleted? The devices cannot send messages, nor can the user view the dashboard – not until the server is restored somehow. If that specific **Availability Zone** (**AZ**) is unreachable temporarily, you would have to manually restore the server to another AZ or Region. If you aren't in a position to do that in an automated fashion, you might be forced to wait until a problem you have no control over is fixed.

- **Scalability** – This architecture might be OK for dozens or even hundreds of devices, and a handful of users polling the Grafana dashboard. However, if demand on either side increases too quickly, the server might well crash under load it is not sized to sustain. Maybe you could fix this by increasing the EC2 instance size for a while – but at a certain number of devices, that is simply not going to work anymore

- You will then need to fundamentally redesign your application to support horizontal scalability across multiple instances. How will you maintain a common state for the dashboard then? How will you balance devices between the instances? These are just the tip of the iceberg of things you now have to worry about.

- **Response time** – What if your application does really well in the market, and you start getting customers from around the world? If this is running in `us-east-1`, will a customer in Australia be happy with the additional 200+ ms latency incurred when they access the dashboard? Will the devices time out trying to get their payloads to the broker

- At this point, you wouldn't just need to scale horizontally – you'd have to do it across multiple geographies. Questions about how you maintain a common state or ensure messages aren't missed or delivered multiple times become far more complex.

- **New features** – Let's say you want to run some kind of analytics on the messages or do some kind of normalization of the data before it reaches the dashboard. You would need to implement additional application components – perhaps a NoSQL database such as MongoDB – and insert it between the existing elements. Probably not an easy change to make.

- **Troubleshooting** – What if there is some difficult-to-pin-down issue such as messages from the devices randomly being lost with no obvious pattern? You will need to implement some form of instrumentation within the application to analyze historical data flows when such problems arise.

- **Compliance** – Assume you wish to service customers in the EU, and the data being gathered is **Personally Identifiable Information** (**PII**). This would make your application subject to GDPR. Now, you need to somehow ensure that the data gathered on EU data subjects is specifically tracked and kept within the country of origin. That is a complicated thing to do if you refuse to use a cloud-native service that is already aware of such things.

- **Security** – How do you handle authentication for the devices or users logging in to the dashboard? A simple local **Privileged Access Management** (**PAM**) database on the server won't scale and will get hard to manage really fast. What if a compliance regime you are subject to insists you identify IoT devices by client-side certificates? How do you distribute them to a fleet of thousands or millions of devices?

- **Cost** – If you wind up deploying EC2 instances around the world, how do you optimize for cost? Native services such as AWS IoT Core charge per message received, but each one of your running EC2 instances charges per hour it's turned on – whether it's being used to its full capacity or not.

Simple IoT application example – the cloud-native version

All of the aforementioned burdens are lightened considerably by using AWS' cloud-native offerings. Instead of building, securing, and scaling your own MQTT broker, you simply leverage the AWS IoT Core endpoints within each region. Traffic can be steered to the best endpoint by pointing to DNS records in Amazon Route 53 with routing by geolocation enabled.

Fleet Provisioning can be used for certificate distribution and device admission. You could easily add new components to the architecture because AWS IoT Core integrates seamlessly with other AWS services such as Lambda, S3, DynamoDB, Kinesis, and more. Features of AWS IoT Core such as Device Shadow would be quite complex to implement yourself.

Finally, dashboards can be built quickly without needing to worry about the infrastructure layer using the Amazon Managed Grafana service. MFA can be delegated to Amazon Cognito for users who log in to the dashboard.

Hopefully, this illustrates that building on EC2 offers more control and can be advantageous in one way; cloud-native services provide significant advantages of their own. These are the same reasons that, in the end, people build applications on the cloud in the first place – which can be summarized as allowing businesses to focus on innovation and less on the underlying nuts and bolts of infrastructure.

Staying ahead of emerging AWS services and trends

Staying ahead of emerging AWS services and trends is crucial for leveraging the full potential of cloud computing. As AWS continuously innovates and evolves, keeping up can be challenging, but it is essential if you want to be ready for the technological breakthroughs expected of Industry 4.0.

Continuous learning

AWS offers a comprehensive `certification program`. More importantly, their training program is top-notch. In addition to traditional classroom-based courses, there is a great deal of self-paced training available in `AWS Skill Builder`:

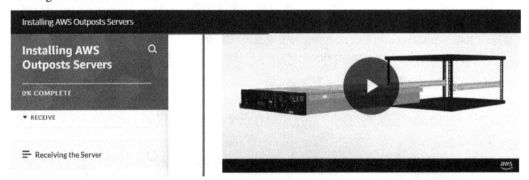

Figure 14.2 – Free training course on AWS Outposts Servers on AWS Skill Builder

Regularly participating in these courses not only keeps you updated on new services and best practices but also deepens your understanding of the AWS ecosystem.

Another type of content available on AWS Skill Builder is hands-on labs that use AWS Workshop Studio to provision you a temporary AWS account to follow along with them:

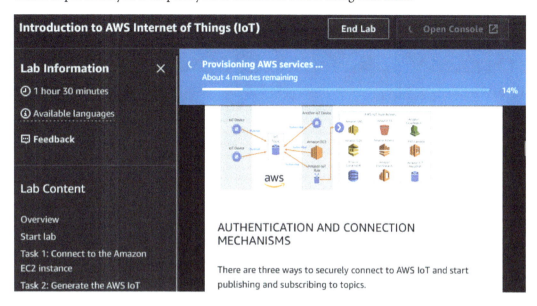

Figure 14.3 – Hands-on lab available in AWS Skill Builder

Note that these labs aren't available with the free tier of the AWS Skill Builder account. A subscription is required. At the time of writing, these subscriptions are $29 a month and grant unlimited access to all of the labs and more advanced training courses in AWS Skill Builder.

Sandbox accounts

Workshop Studio labs are great, but keep in mind they are tightly restricted to the services involved in the self-paced learning module associated with them. You couldn't, for example, use a Workshop Studio lab about DynamoDB to build an IoT application. SCPs are in place to prevent you from using services or features outside the scope of the lab.

This is why you should, as part of your multi-account strategy, set aside one or more accounts for the explicit purpose of being a sandbox that your team can use to play with new services, without needing to worry about impacting any of your production or dev/test/qa stages. The tricky part with this is ensuring resources deployed for "playing around" don't get accidentally left running and incur charges after such activities are complete.

In such cases, AWS Labs publishes a solution on GitHub called `Sandbox Accounts for Events`. It is based on Optum's open source `Disposable Cloud Environment (DCE)` project. The Sandbox Accounts for Events solution allows you to provide multiple temporary AWS accounts that are automatically deleted after a period of time you specify:

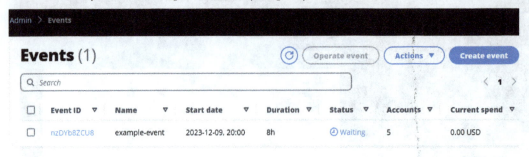

Figure 14.4 – Using the Sandbox Accounts for Events solution from AWS Labs

It uses the concept of leases to define the length of time an account can live, as well as the maximum spend limits per account. For instance, you could assign a temporary account to a new hire on your team for a 1-week period and set a maximum spend limit of $100. If either of those conditions are met, the account and any resources inside are deleted.

Summary

In this chapter, we reviewed what Industry 4.0 is and why it is at the forefront of public and private organizations' technology strategies. We covered the benefits it is predicted to yield, as well as the challenges it is likely to present as the demand for such applications scales exponentially in accordance with expected developments in near-future device technology.

Next, we shifted to a discussion of things you should take into account now as you architect and deploy your distributed edge computing applications to ensure they are ready for future developments, both within the industry and AWS in particular. This includes leveraging AWS Cloud Foundations and the AWS Well-Architected Framework.

Lastly, we covered some recommended best practices as well as pitfalls to avoid as you get started building your edge compute applications on AWS. This included examples of advantages and disadvantages of the types of AWS services you choose to build with, as well as strategies to stay abreast of emerging AWS services as they are released.

Index

www.packtpub.com

Subscribe to our online digital library for full access to over 7,000 books and videos, as well as industry leading tools to help you plan your personal development and advance your career. For more information, please visit our website.

Why subscribe?

- Spend less time learning and more time coding with practical eBooks and Videos from over 4,000 industry professionals

- Improve your learning with Skill Plans built especially for you

- Get a free eBook or video every month

- Fully searchable for easy access to vital information

- Copy and paste, print, and bookmark content

Did you know that Packt offers eBook versions of every book published, with PDF and ePub files available? You can upgrade to the eBook version at packtpub.com and as a print book customer, you are entitled to a discount on the eBook copy. Get in touch with us at customercare@packtpub.com for more details.

At www.packtpub.com, you can also read a collection of free technical articles, sign up for a range of free newsletters, and receive exclusive discounts and offers on Packt books and eBooks.

Other Books You May Enjoy

If you enjoyed this book, you may be interested in these other books by Packt:

AWS Observability Handbook

Phani Kumar Lingamallu | Fabio Braga de Oliveira

ISBN: 978-1-80461-671-0

- Capture metrics from an EC2 instance and visualize them on a dashboard
- Conduct distributed tracing using AWS X-Ray
- Derive operational metrics and set up alerting using CloudWatch
- Achieve observability of containerized applications in ECS and EKS
- Explore the practical implementation of observability for AWS Lambda
- Observe your applications using Amazon managed Prometheus, Grafana, and OpenSearch services
- Gain insights into operational data using ML services on AWS
- Understand the role of observability in the cloud adoption framework

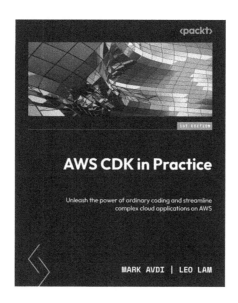

AWS CDK in Practice

Mark Avdi | Leo Lam

ISBN: 978-1-80181-239-9

- Turn containerized web applications into fully managed solutions
- Explore the benefits of building DevOps into everyday code with AWS CDK
- Uncover the potential of AWS services with CDK
- Create a serverless-focused local development environment
- Self-assemble projects with CI/CD and automated live testing
- Build the complete path from development to production with AWS CDK
- Become well versed in dealing with production issues through best practices

Packt is searching for authors like you

If you're interested in becoming an author for Packt, please visit `authors.packtpub.com` and apply today. We have worked with thousands of developers and tech professionals, just like you, to help them share their insight with the global tech community. You can make a general application, apply for a specific hot topic that we are recruiting an author for, or submit your own idea.

Share Your Thoughts

Now you've finished *Edge Computing with Amazon Web Services*, we'd love to hear your thoughts! Scan the QR code below to go straight to the Amazon review page for this book and share your feedback or leave a review on the site that you purchased it from.

`https://packt.link/r/1835081088`

Your review is important to us and the tech community and will help us make sure we're delivering excellent quality content.

Download a free PDF copy of this book

Thanks for purchasing this book!

Do you like to read on the go but are unable to carry your print books everywhere?

Is your eBook purchase not compatible with the device of your choice?

Don't worry, now with every Packt book you get a DRM-free PDF version of that book at no cost.

Read anywhere, any place, on any device. Search, copy, and paste code from your favorite technical books directly into your application.

The perks don't stop there, you can get exclusive access to discounts, newsletters, and great free content in your inbox daily

Follow these simple steps to get the benefits:

1. Scan the QR code or visit the link below

https://packt.link/free-ebook/9781835081082

2. Submit your proof of purchase
3. That's it! We'll send your free PDF and other benefits to your email directly

Download a free copy of this book

www.ingramcontent.com/pod-product-compliance
Lightning Source LLC
Chambersburg PA
CBHW080612060326
40690CB00021B/4667